Acknowledgments

This curriculum was developed through funds from a USDA-NIFA Grant (Award # 2012-68003-19621) titled "Development of Novel Salmonella Control Practices and Integrated Education Program to Reduce Salmonellosis." Research **Principal Investigators** include Dr. Hosni Hassan, Dr. Matthew Koci at North Carolina State University, and Dr. Andrea Azcarate-Peril at the University of North Carolina at Chapel Hill. Extension and education components for this grant were provided by North Carolina 4-H and the North Carolina Kenan Fellows Program, both at North Carolina State University.

Also instrumental in the creation of this curriculum:

Kenan Fellow Authors
Victoria Raymond
Whitney Baker
Kalyani Tawade

4-H Curriculum Specialist
Amy Chilcote

4-H Evaluation Specialist
Autumn Guin

4-H Microbiology Program Coordinator
Anne Greenaway

4-H Graphic Design/Illustration
Tabitha Barnes

4-H Illustrator
Uyen Nguyen- NCSU Chemistry Undergraduate

4-H curriculum is vetted through both formal and non-formal educational environments. This curriculum has been piloted through the NC 4-H Curriculum Guidelines and evaluated through the following youth directed systems: summer camp, school enrichment, and 4-H community clubs. The following counties assisted in piloting this curriculum: Cabarrus, Cherokee, Johnston, Rowan, Sampson, and Wilson County Cooperative Extensions.

Introduction

The unseen world of bacteria, the microscopic kingdom of life, has an enormous impact on our everyday lives. This series of lessons and activities seeks to provide participants with an interactive look at the microbial world from a variety of approaches.

First, participants are introduced to the amazing breadth of this group as they become familiar with the distinguishing characteristics of some representative bacteria that are commonly found in association with humans. Then participants are introduced to the increasingly important concept of the microbiome – the populations of bacteria that inhabit all of our bodies and that contribute to health as well as disease.

Next, learners investigate the mechanisms by which the human body keeps any harmful invaders at bay, both by external protective means as well as through cellular armies that attack pathogens.

The importance of public health organizations and surveillance is demonstrated as learners discover how infectious diseases are tracked and how epidemiologists function to control outbreaks.

Participants then step into the operational world by modeling several tools currently available to scientists for identifying bacteria at the molecular level. Participants model gel electrophoresis of DNA molecules as well as DNA sequencing strategies that are on the cutting edge of research.

Environmental factors that can affect the growth of bacterial populations are explored in the next lesson. These tiny organisms reproduce very quickly, which can give them an edge when it comes to growing large populations very fast.

Finally, participants can practice several essential lab techniques that are mainstays for anyone seeking to work in a research or lab environment. These hands-on skills are highly desirable in today's technical workplace. Careers in microbiology can become more appealing through participation in the curriculum.

ADDITIONAL MATERIALS

There are additional materials for each lesson available on our NC 4-H curriculum website (nc4hcurriculum.org).

Content for lessons 1-6:

- Lesson variations for remote learning
- Videos
- PowerPoint presentations
- Interactive content
- PDFs separated by activity for easy printing
- Fillable PDFs for online submission of worksheets
- New content added regularly

Visit *NC4HCurriculum.org* under the Virtual Resources menu section.

Bonus access to all "Grab and Go" content on the menu under Grab and Go. Those are not lessons from this microbiology curriculum.

Access Code for "*Microbes: The World Within*" and all "*Grab and Go*" content:

60b101c9c78c7

Table of Contents

Table of Contents

Scope and Sequence

High School Microbiology Unit
Scope and Sequence Variations

Teaching Lessons over 5 days for 1 ½-2 hours a day:

DAY	LESSONS
1	Lesson 1: Do These Germs Make Me Look Fat? Grab and Go 2: Plate 'em Up (if time permits, begin the plating activity)
2	Lesson 2A: Dare to Infect Me! Lesson 2B: Code Red! (begin banana learning experience)
3	Finish 2B as necessary Lesson 3: Outbreak
4	Lesson 5: 1, 2, 4, Grow
5	Lesson 6: Double, Double, Toil, and Trouble

Teaching Lessons over 3 days: 2-3 hours a day:

DAY	LESSONS
1	Lesson 2A: Dare to Infect Me! (begin banana learning experience) Lesson 1: Do These Germs Make Me Look Fat?
2	Lesson 2B: Code Red! Lesson 3: Outbreak
3	Lesson 5: 1, 2, 4, Grow Lesson 6: Double, Double, Toil, and Trouble

Teaching all lessons in 1 day: 6-7 hours:

DAY	LESSONS
1	Lesson 1: Do These Germs Make Me Look Fat? Lesson 2A: Dare to Infect Me! (show the banana video) Lesson 2B: Coe Red! Lesson 5: 1, 2, 4, Grow Lesson 6: Double, Double, Toil, and Trouble

Experiential Learning: A Holistic Approach to Learning

Experiential Learning is student centered. It is an active learning process, which first involves youth in an activity, asks them to reflect on the activity to determine what was useful or important to remember, and then tasks youth with understanding how the information they learned applies to their lives now and in the future. This learning process is holistic and allows youth to consume, apply, and build upon life experiences.

Pfeiffer and Jones' experiential learning model is commonly used in 4-H. However, other pioneers such as Kolb provide a wealth of research that emphasizes the importance of the adult facilitator. According to Kolb, the experiential learning model allows the learner actively participate in what is learned. The facilitator's role is to connect with the student to spark their interest and intrinsic motivation so that the student can gain, revise, or enhance their life skills. The facilitator plays multiple rolls during this process: balancing the attention of the learner, engaging with the subject matter, and creating a safe and holistic learning environment.

Tips to think about when creating a welcoming environment for learning:

- Patience is a virtue. When waiting for youth to respond fully, having patience and allowing time for them to think is important.
- You are not the judge. Accept all responses.
- Experiential questions are open-ended to allow for expression. Make sure that you acknowledge youth who respond to you. Not acknowledging all youth can stop the flow of answers and make youth less likely to respond in the future.
- You are not the young person. Don't answer the question.
- Encourage deep conversation. If a young person gives a short response, provide follow up questions to allow for deeper conversation. Examples could be: "Would you provide an example of what you mean? Can you tell me more about your thoughts?"
- Avoid yes or no questions. These are not open-ended and stop the flow of conversation.
- Don't lead the youth to answers you want to hear.
- Allow the youth to share their responses in pairs rather than with the entire group.

1. Experience: Let's Do It!!!
   Key Reminders to Leader:
- "Sit on Your Hands"
- Observe
- Facilitate to the "bigger picture"

Key Objectives for Discovery:
- To Explore
- To Examine
- To Construct
- To Arrange

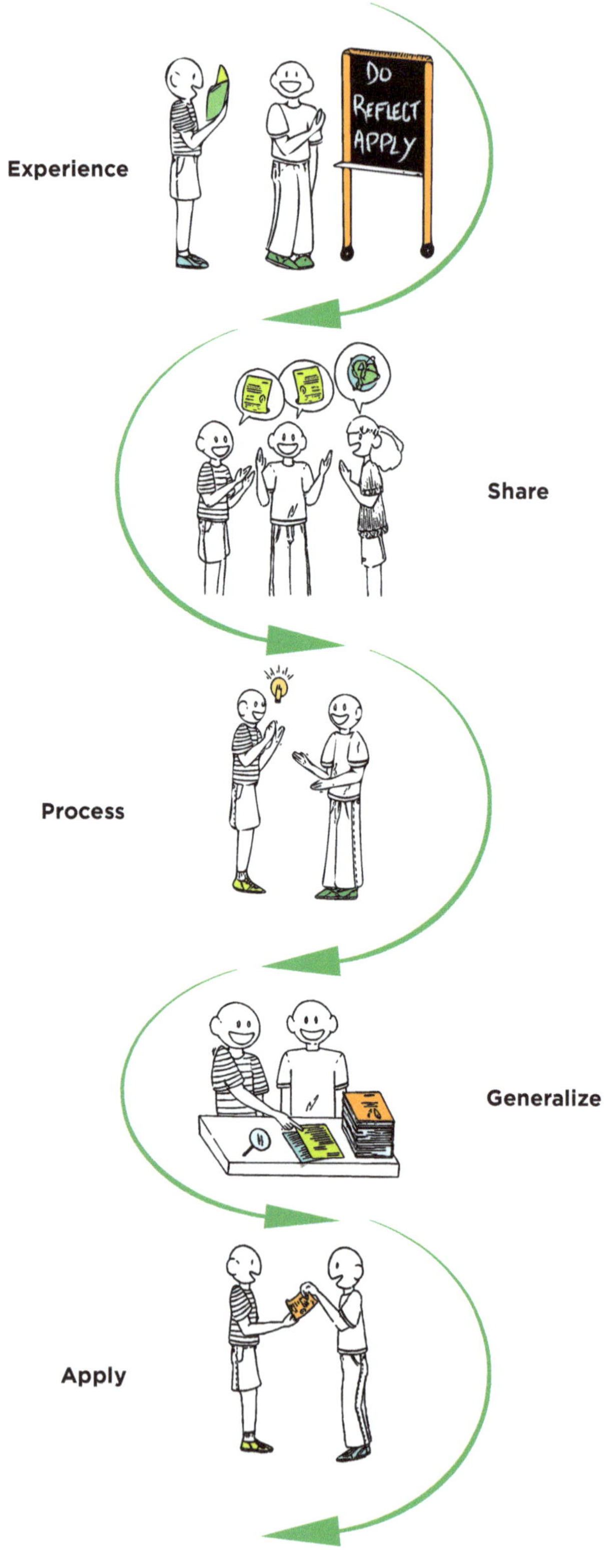

2. Share
Key Concept: Respond
Key Question: "What Happened?"

3. Reflect
Key Concept: Analyze Patterns
Key Question: "What's Important?"

Leader Role
- Allow adequate processing time for sharing.
- Use open-ended questions to stimulate thinking and feeling.
- Encourage "pair share" and large group share.

4. Generalize
Key Concept: Inference
Key Question: "So What?"

5. Apply
Key Concept: Application
Key Question: "Now What?"

Leader Role
- Facilitate youth finding ways to use what they have learned in new situations.
- Guide youth in making connections between the inner meaning of the activity and the broader world.

Share Questions (What Happened?)
What did you like about this activity?
What part of the activity was the hardest to do?
How did you decide what to choose?
What was the most fun about doing the activity?
How did your group work together?
How did you feel about successfully being able to____?
What are some decisions you had to make to carry out this activity?

What steps did you go through before you made your decision?
Why do you think everyone in your group didn't agree about everything?
What was it like to have to make quick decisions?
How did you keep track of everyone's ideas?
Do you think you get more ideas working alone or in a group? Why?
What kind of feelings did you have when group members argued?
What did you observe about the way the groups disagreed (or agreed)?
How did you use your various senses to ___?
What new words did you learn?
Why do you think people have different ideas about what is correct?

Reflect Questions (What's Important?)
What did you learn about yourself by doing this activity?
Why is it important to know___?
How did your group decide ___?
What did you learn as a group that you might not have learned alone?
What were some common themes or thoughts you heard?
How were each person's viewpoints the same? How were they different?
What problems came up over and over?
What would you do if_____ were to happen?
What was easy or difficult about working with a group to____?
What did you do if everyone in the group didn't agree on___?
How did this differ from the way you are usually taught in school?
Give an example of a challenge you had and what you did to solve it.
What works best to get people involved and excited about doing this type of activity?
Why is learning with others sometimes more fun than learning alone?
What suggestions would you have for someone who wanted to____?

Generalize Questions (So What?)
What other situations like this have you experienced?
Why is it important that each person has his/her own view?
When have you had to___ before?
Where can you find resources to help you make some of your decisions about__?
When else have you had fun and learned new things at the same time?
Why is it important to have plenty of information before making decisions?
When do you make decisions that require everyone in the group to agree?
What do you do when you don't agree with the group?
What did you learn about your own skill in making decisions?
What did you learn about your own skill in communicating with others?
Describe five ways in which new ideas can be communicated to you.
In what ways do people help each other learn new things?
What are some ways you like to learn?
What did you learn?

Apply (What was Learned?)
How would you teach someone about this activity or concept?
What did you learn today that you will be able to use in school....at home?
What did you learn by participating in this process that will help you in the future?
Describe a time when you might need the skills/knowledge you learned today.
What are some other situations when you will need to use the skills you learned today?
What would you do differently if you conducted this activity again?

LESSON 1 DO THESE GERMS MAKE ME LOOK FAT?

THE ROLE OF THE MICROBIOME IN HEALTH AND DISEASE

Introduction

Bacteria are single-celled prokaryotic organisms. They represent very diverse species. Bacteria are distinguished by their shape, their ability to move (and structures that assist in that ability), their oxygen requirements (aerobic [needing oxygen] and anaerobic [not needing oxygen] or both), their particular nutrient needs, and their reactions to chemical stains.

Bacteria in a given niche create a community of interacting organisms called a microbiome. This community can live on and in multicellular organisms such as humans. The majority of the members of the microbiome cause no harm to the human but in fact, provide helpful functions. To understand the variety and complexity of the microbiome, the participant must become familiar with some representative characteristics and examples of microbial life.

SALMONELLA

Gram negative
Aerobic/anaerobic
Rod-shaped
Motile
No spores
Cannot ferment lactose

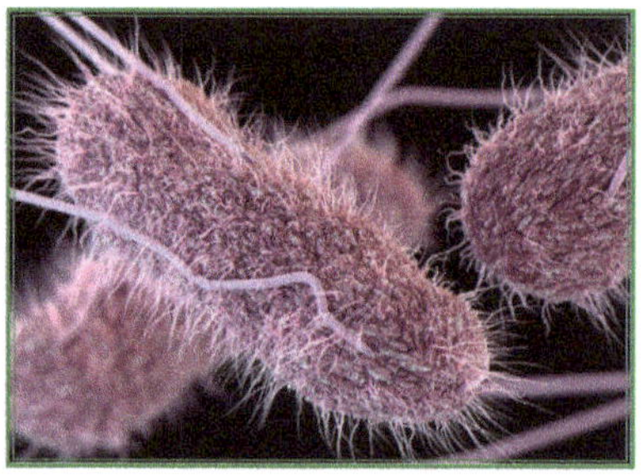

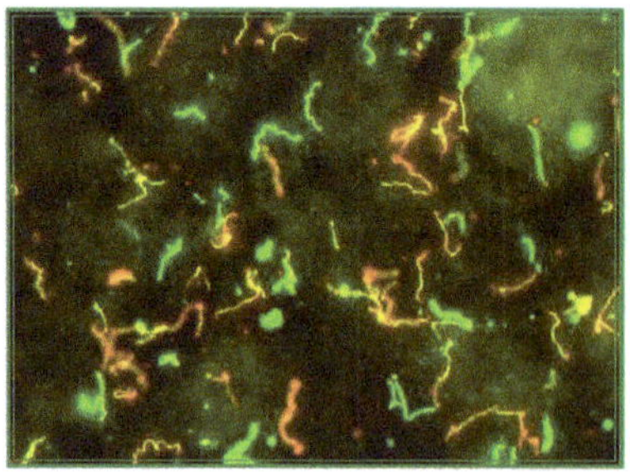

BORRELIA

Gram negative
Anaerobic
Spiral-shaped
Motile
No spores

STREPTOCOCCUS

Gram positive
Aerobic/anaerobic
Spherical
Non-motile
No spores
Occur in strings

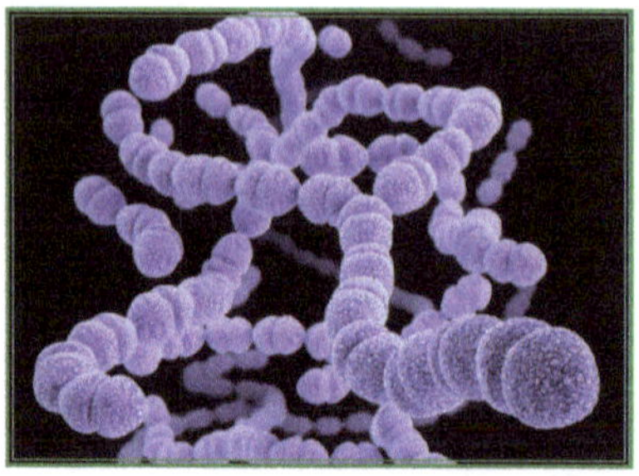

LESSON 1 DO THESE GERMS MAKE ME LOOK FAT?

Key Teaching Points

1. Bacteria are prokaryotic organisms that are classified by shape, by how they move, by their need for oxygen, and by their response to stain.
2. Bacteria make up a large percentage of the biomass of the earth.
3. Bacteria play important roles in the world, including decomposition, food production, and the human microbiome.

Learner Outcomes

- Recognize the presence of microscopic life all around us.
- Distinguish the difference between microbial life and multicellular organisms.
- Construct an understanding of the diversity of microbial organisms.

Skill Level: Advanced, ages 14-18

Success Indicators

Participants will develop an organizational scheme for a diverse group of microbes.

- Participants will create an argument poster that presents the main ideas in an informative text.
- Participants will demonstrate a new awareness of the importance of microscopic life on and in the body.

Life Skills

Communication, Cooperation/Teamwork, Planning/Organizing

Time Needed

Activity 1: Microbes all Around Us (30 minutes)
Activity 2: Microbes and Me (50 minutes)

Space

Meeting room

Suggested Group Size

Groups of 4

Materials List

- Bacteria Fortune Teller (1 per participant)
- Set of 12 Bacterial Mug Shots (1 set/group of 4)
- Dichotomous Key (1 page/group of 4)
- Dichotomous Key Example (for facilitator use)
- "Body Bacteria: Exploring the Skin's Microbial Metropolis" article (1 copy/participant) https://bit.ly/3piEFTD
- Poster paper or clean 8 ½" x 11" paper
- Markers, scissors

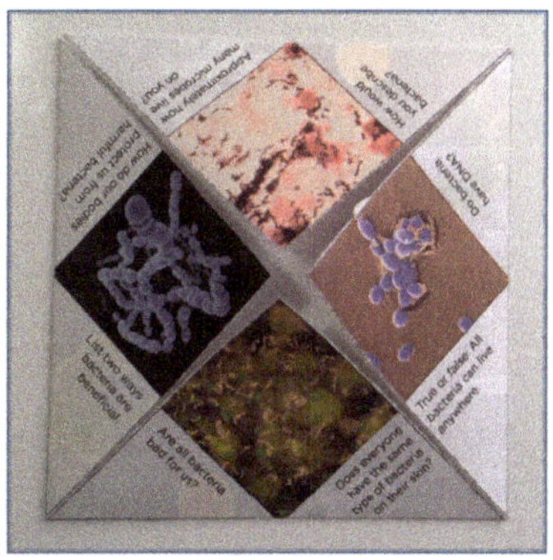

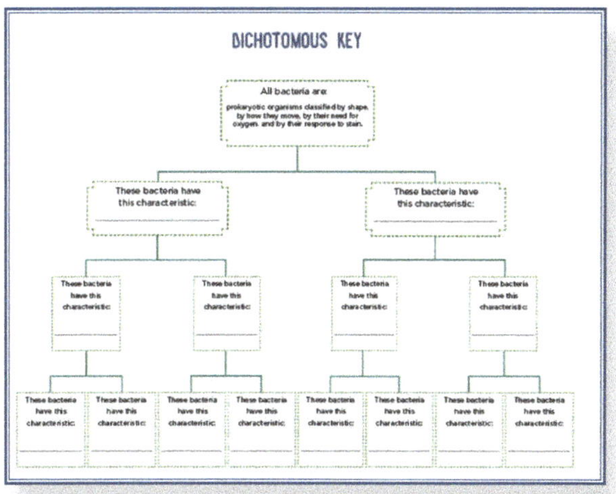

Before the Activity

1. Make copies of the Bacteria Fortune Teller.
2. Make copies of the Dichotomous Key.
3. Make a copy of the Dichotomous Key Example for facilitator use.
4. Review the presentation for this lesson.
5. Prepare technology to show the video and presentation to the participants.
6. Read the article "Body Bacteria" to become familiar with the information.

Activity 1: Microbes All Around Us!

Opening Question

How much of you is bacteria?

How much do YOU know about bacteria?

1. Participants will construct a Bacteria Fortune Teller according to the folding instructions.

2. In pairs, participants will quiz each other on questions in the Fortune Teller to assess prior knowledge.

3. After 10 minutes, participants will share any new insights that they have made with the entire group.

4. Present the "Bacterial Diversity and the Microbiome" presentation to the participants.

5. Included as video link in the presentation:
"We're Mainly Microbe" from PBS (3:50 minutes)
https://youtu.be/4BZME8H7-KU

Who are some of these bacteria?

1. Separate participants into groups of 3-4. Distribute Bacteria Mug Shots (1 set of 12 to each working group). Note information contained on cards. Distribute Dichotomous Key (1 per working group).

2. Participants organize their set of Mug Shots according to similar characteristics. To get started, participants will look at all the mug shots in the set and organize them into two groups based on one characteristic. They will then sort the mug shots into two piles. Participants will then examine each of the two new groups and divide them again. The groups do not have to divide evenly - there may be

a division that assigns three bacteria to a group and one to another. Continue for three levels of organization. Participants will record their group's scheme on their Dichotomous Key.

3. COMPARE one group's scheme with another group's. If there is only one group, SUGGEST an alternate way that the groups could be divided.

4. EXCHANGE one group's key with another to see if they can work successfully to identify the bacteria.

5. The facilitator asks the entire group:
   How do these varied bacteria relate to us as humans?
   Did you recognize any of the bacterial names from the video or TV ads or news?

Activity 2: Microbes and Me

"Body Bacteria" article Jigsaw

1. Divide the participants into 5 groups:

   a. The Good, the Bad, and the Acne/Belly Button Bacteria
   b. Diverse Settlements
   c. Microbes in Medicine/A Problem Afoot
   d. Defense Mechanisms
   e. More than Skin Deep

2. Each group reads the indicated portion of the article. In the group, answer the following questions to summarize the section for the remaining participants: (10 minutes)

   a. What is the main point of the section?
   b. What evidence does the writer use?
   c. What specialized terms or knowledge do you need to understand the section?
   d. What visual might be used to represent the section?

3. Re-arrange the participants into new groups. Each group will contain one participant from each of the original groups. Participants will report to the new group about their portion of the article. Groups will begin planning their poster.

4. Create a poster presenting the microbiome either using large paper or by using http://www.easel.ly/ or http://infogr.am/ (30 minutes).

- The poster should contain the title "Body Bacteria"
- The poster should have at least 3 interesting visual elements
- The poster should contain elements contributed from each of the 1-5 groups

5. Groups will present their posters to the entire class in a walking gallery. Each group will display their poster on a wall or using appropriate technology and all participants will rotate to visit each poster.

6. Ask the Share question.

    - Do any of the posters present a different emphasis than yours?

7. Ask the Reflect question.

    - What argument for bacteria as "good guys" can you now make that you might not have been able to make before?

8. Ask the Generalize question.

    - How might people benefit from knowing more about their microbiome?

9. Watch the YouTube video "Bacteria in gut may cause morbid obesity" from CNN (1:30 minutes)
https://youtu.be/q4LtSJoSPpA

10. Ask the Apply question.

    - How do you think obesity treatment will differ in the future as more information about the microbiome becomes available?

Talk It Over

SHARE 1	REFLECT 2
Do any of the posters present a different emphasis than yours?	What argument for bacteria as "good guys" can you now make that you might not have been able to make before?
GENERALIZE 3	**APPLY 4**
How might people benefit from knowing more about their microbiome?	How do you think treatment for obesity will differ in the future as more information about the microbiome becomes available?

Lesson 1 ✱ Do These Germs Make Me Look Fat?

Term and Concept Discovery

Biodiversity: the variety of species that inhabit an ecosystem

Bacteria: a type of single-celled organism. Bacteria are classified as prokaryotic organisms, as they do not have membrane-enclosed nuclei or other similar structures (organelles). Bacteria are the most abundant life forms on our planet. Singular: bacterium

Microbiome: the term given to the microbes that live within/on a particular organism

Prokaryote: Any cellular organism that has no nuclear membrane, no organelles in the cytoplasm except ribosomes, and has its genetic material in the form of single continuous strands forming coils or loops, characteristic of all organisms in the kingdom Monera, as the bacteria and blue-green algae.

Variations

The entire "Body Bacteria" article may be read by each participant and a poster generated separately to be shared at the next meeting.

Learn More

Zit Zapper: How (Good) Bacteria Could Be the Answer to Clear Skin
https://bit.ly/3uUU91C

Understanding the Good Bacteria in Our Skin
https://youtu.be/LvVThKy7FCQ

After weight-loss surgery, new gut bacteria keep obesity away. https://reut.rs/3fLZP9x

"The Gut Microbiome Influences Whether or Not You're Fat"
https://bit.ly/3uOVFCn

The Bellybutton Biodiversity Project http://robdunnlab.com/projects/belly-button-biodiversity/

Education Standard(s)

NGSS HS-LS 2-6 Evaluate the claims, evidence, and reasoning that the complex interactions in ecosystems maintain relatively consistent numbers and types of organisms in stable conditions, but changing conditions may result in a new ecosystem.
CCSS.ELA-Literacy.RST.9-10.2 Determine the central ideas or conclusions of a text; trace the text's explanation or depiction of a complex process, phenomenon, or concept; provide an accurate summary of the text.

Tags

bacteria, microbiome, habitat, microbes, biology

Lesson 1 Print Materials

Facilitator Print Materials
Dichotomous Key Example

Group Handouts
Dichotomous Key

Youth Handouts
Bacteria Mug Shots
Bacteria Fortune Teller
"Body Bacteria: Exploring the Skin's Microbial Metropolis" article. Print from this link: https://bit.ly/3piEFTD

See page 3 for access to digital assets & extra enrichment.

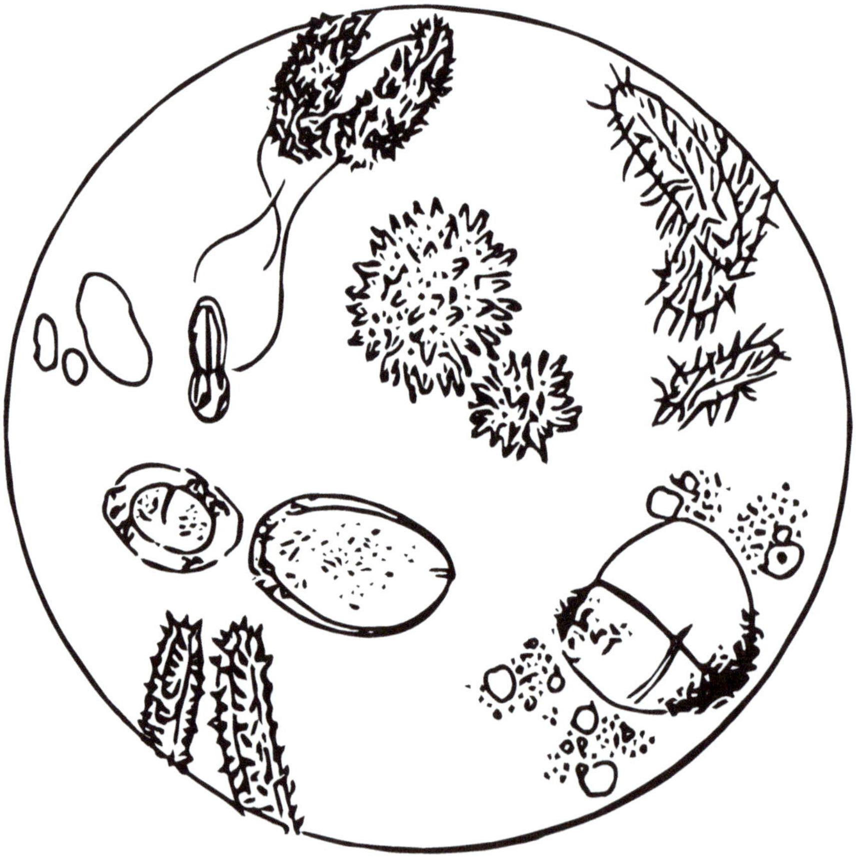

Bacteria are diverse, prokaryotic organisms that are classified by their shape, movement, oxygen needs, nutrient needs, and response to stain. Microbes make up a large percentage of the Earth's biomass and are responsible for various roles within a microbiome. In this lesson, you will learn to recognize the presence of microscopic life around us.

DICHOTOMOUS KEY EXAMPLE

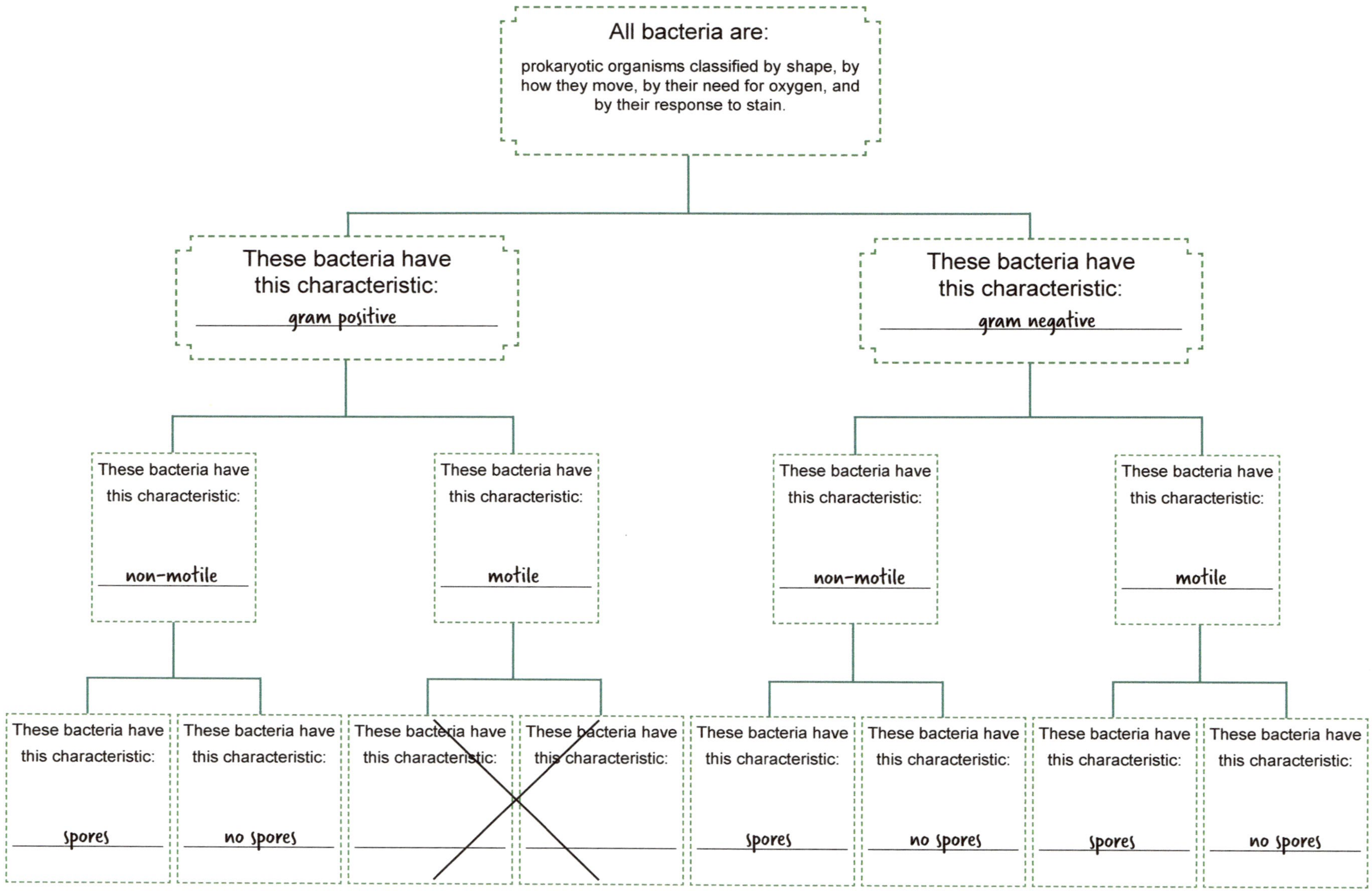

DICHOTOMOUS KEY

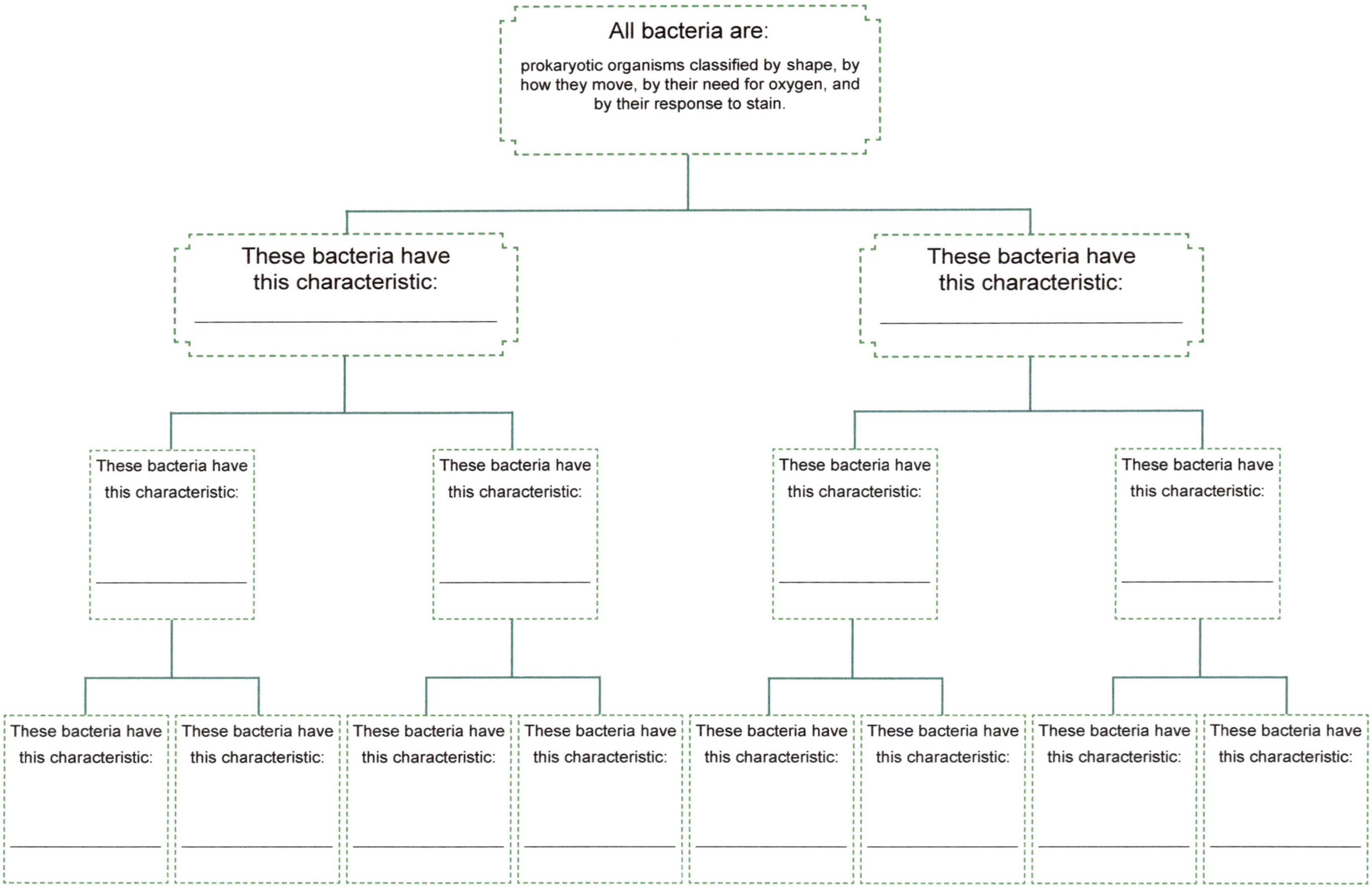

Bacteria Mug Shots

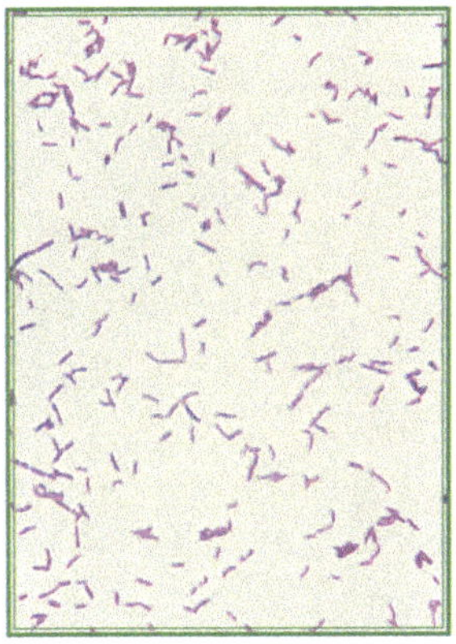

Lactobacillus
Gram positive
Aerobic/anaerobic
Rod-shaped
Non-motile
No spores

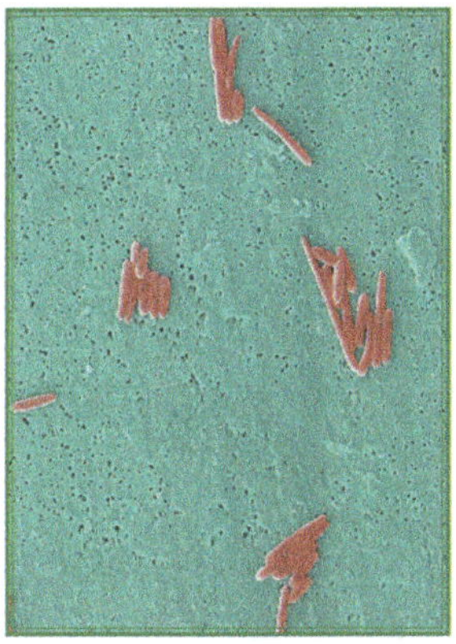

Bacteroides
Gram negative
Anaerobic
Rod-shaped
Non-motile
No spores

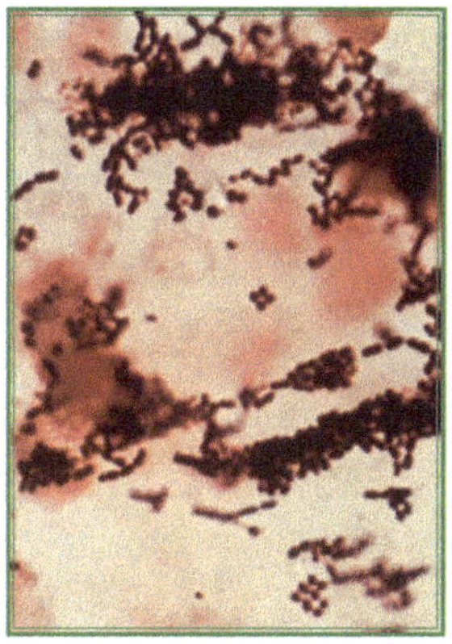

Peptostreptococcus
Gram positive
Anaerobic
Spherical
Non-motile
No spores
Occur in strings

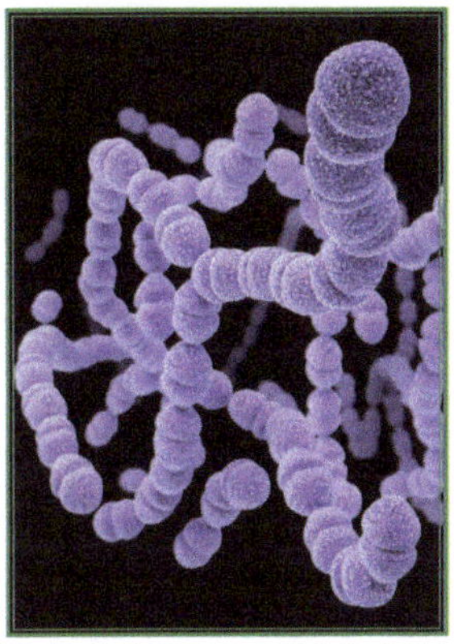

Streptococcus
Gram positive
Aerobic/anaerobic
Spherical
Non-motile
No spores
Occur in strings

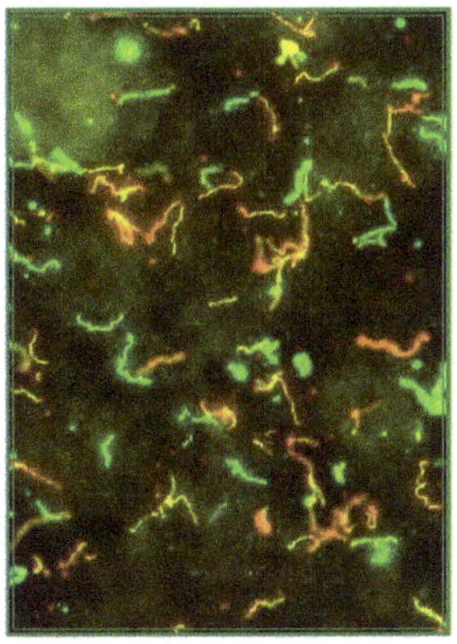

Borrelia
Gram negative
Anaerobic
Spiral-shaped
Motile
No spores

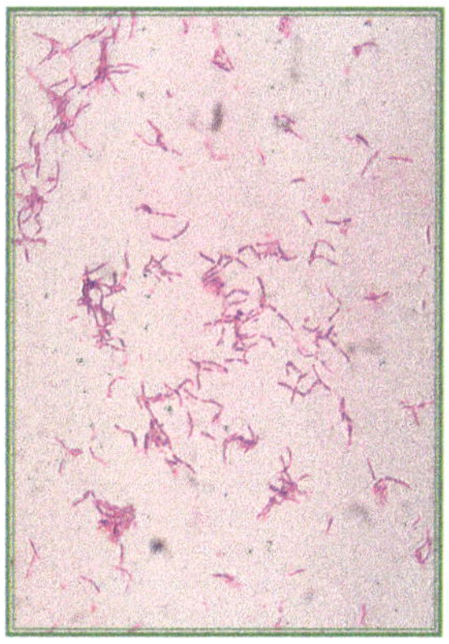

Bifidobacterium
Gram positive
Anaerobic
Rod-shaped
Non-motile
No spores

Bacteria Mug Shots

Salmonella

Gram negative
Aerobic/anaerobic
Rod-shaped
Motile
No spores
Cannot ferment lactose

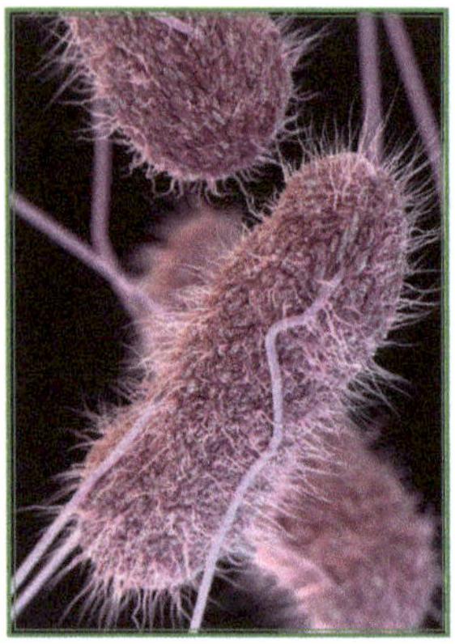

Enterococcus

Gram positive
Aerobic/anaerobic
Spherical pair
Non-motile
No spores

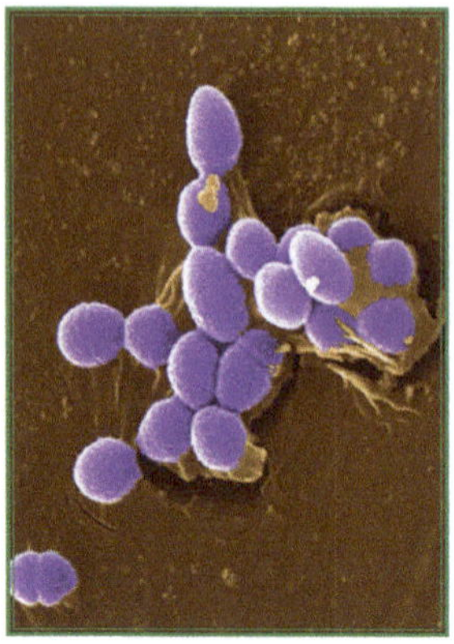

Clostridium

Gram positive
Anaerobic
Rod-shaped
Non-motile
Spores

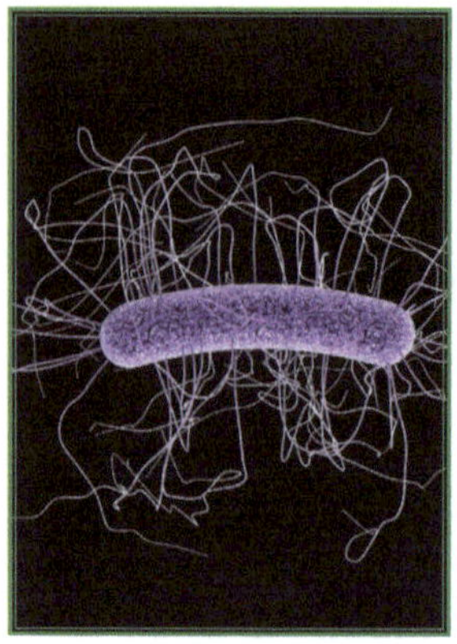

Helicobacter

Gram negative
Aerobic
Spiral-shaped
Motile
No spores

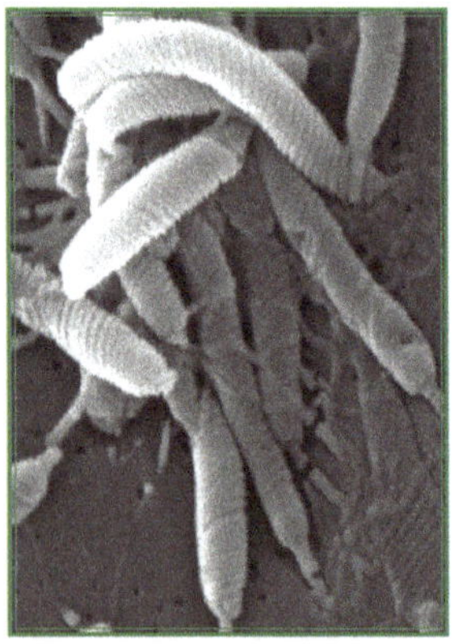

E. Coli

Gram negative
Aerobic/anaerobic
Rod-shaped
Motile
No spores
Can ferment lactose

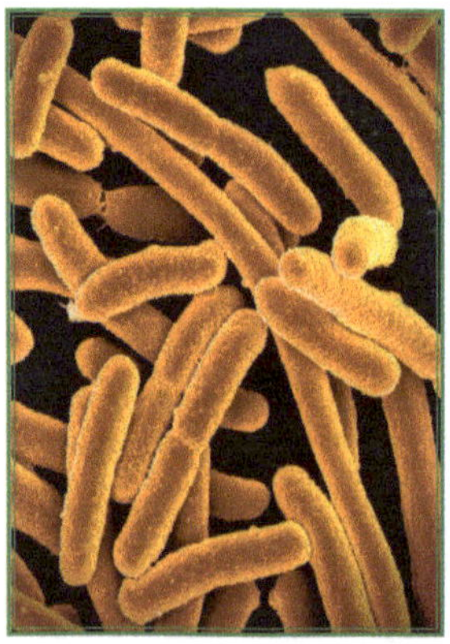

Staphylococcus

Gram positive
Aerobic/anaerobic
Spherical
Non-motile
No spores
Occur in clumps

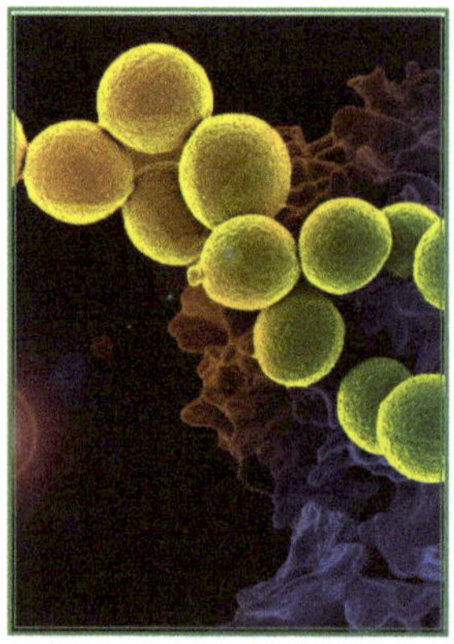

Bacteria Fortune Teller

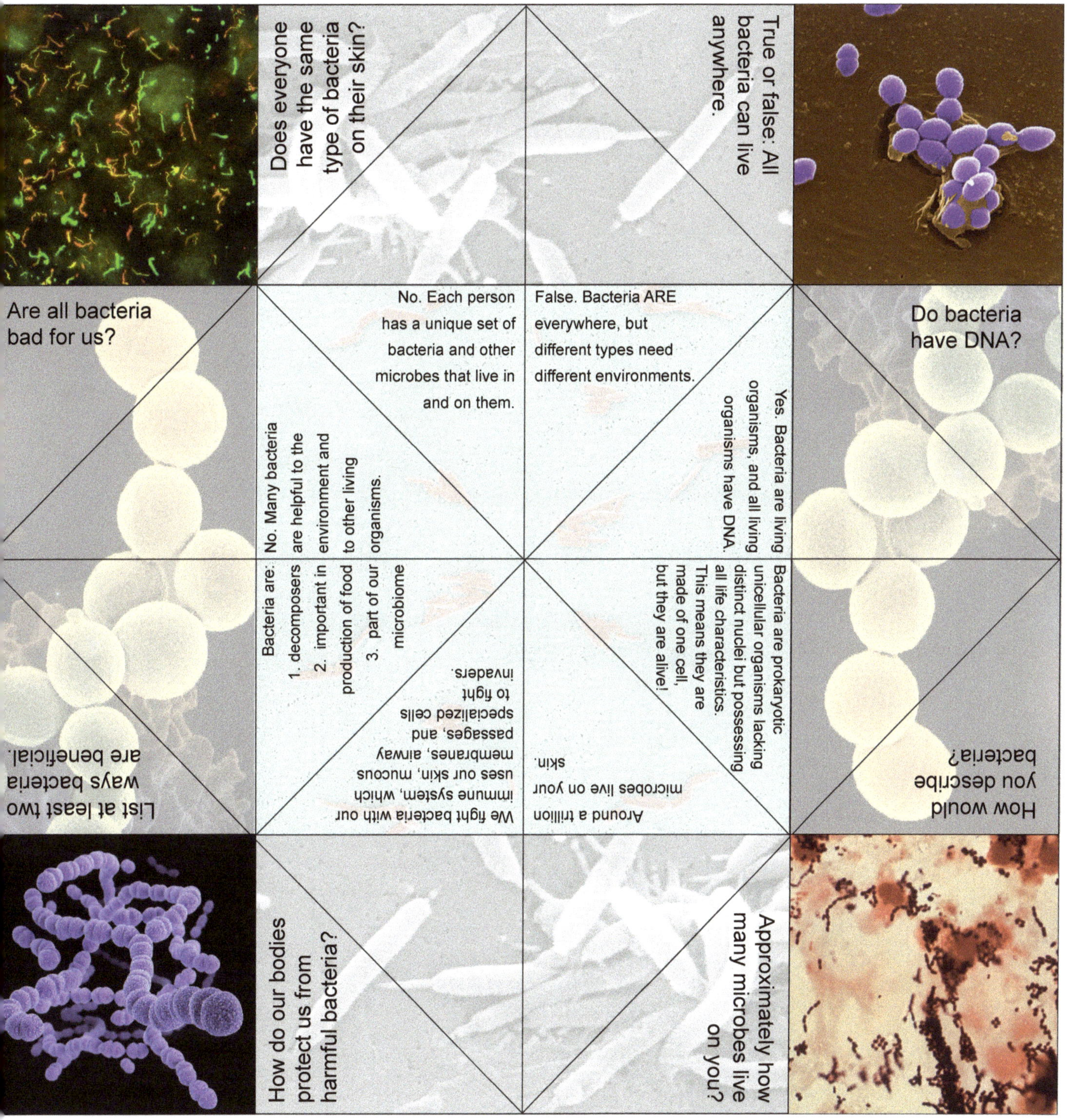

Bacteria Fortune Teller Instructions

1. Cut out the fortune teller.

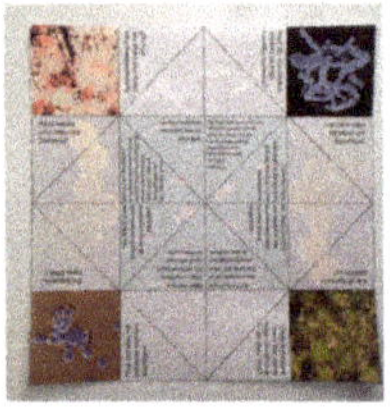

2. Lay upside down.

3. Fold all corners to the center.

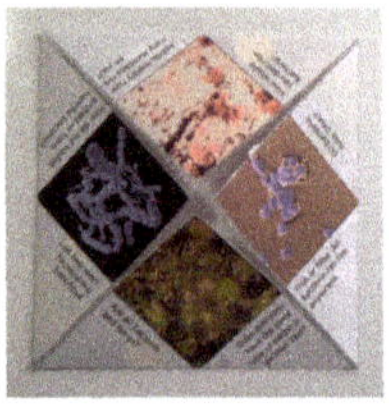

4. Turn over and fold all corners to the center again.

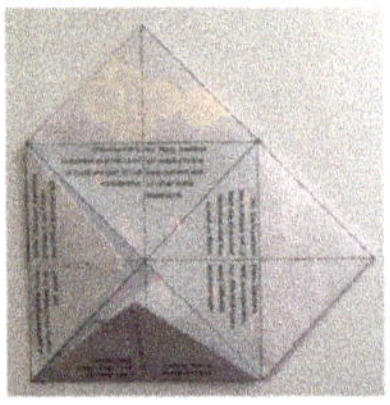

5. Fold in half twice.

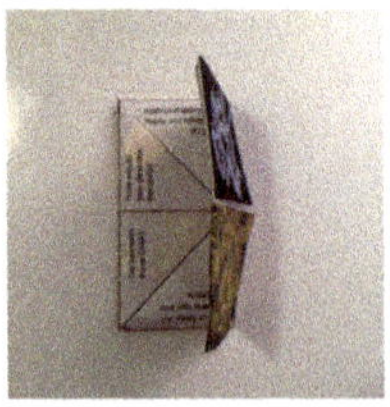

6. Insert fingers to play.

LESSON 2A DARE TO INFECT ME

HOW THE HUMAN BODY IS PROTECTED FROM PATHOGENS

Introduction

What is an immune System?

The **immune system** is a network of cells, tissues, and organs that protects the human body. The defense mechanisms of the immune system can be divided into two types.

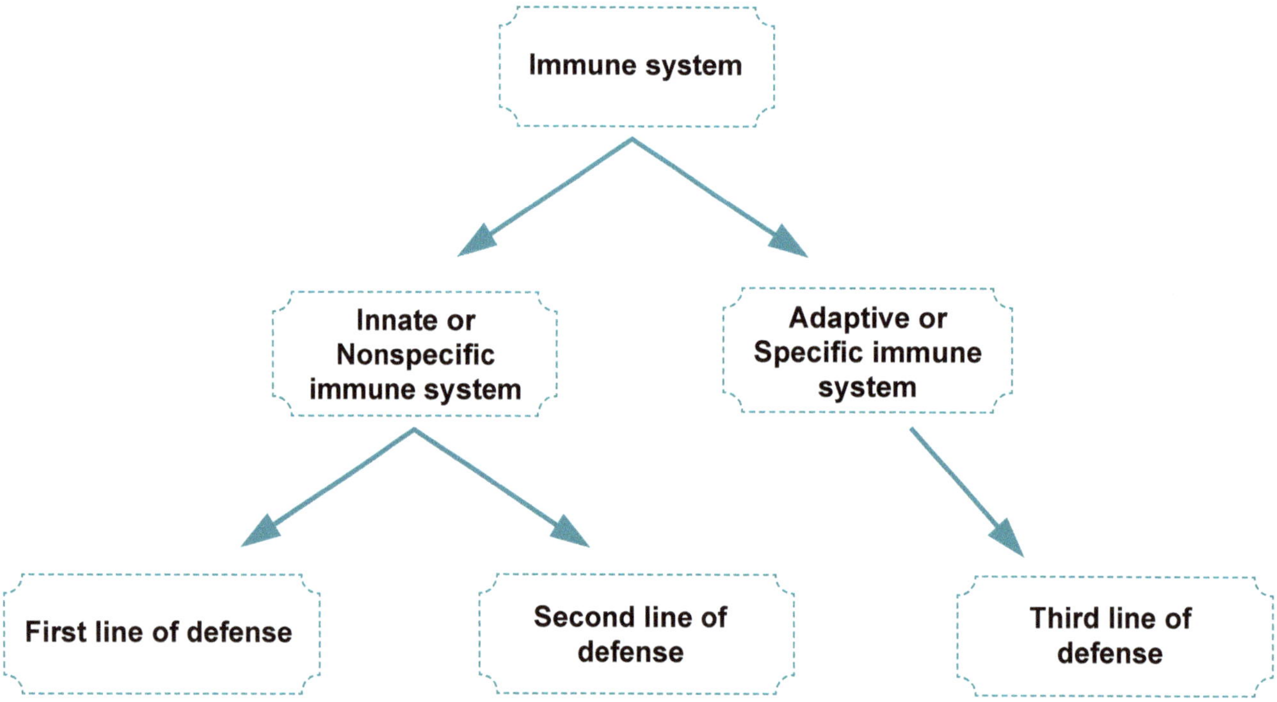

Innate immunity is nonspecific immunity and works against wide variety of invasive organisms. The innate immunity system includes the First line of defense and the Second line of defense. The first line of defense includes barriers such as the skin, tears, mucus, and low pH, which are toxic to or destroy invaders. The second line of defense includes phagocytic cells which ingest foreign cells and other particles.

Adaptive immunity is specific. It distinguishes between substances made by the organism (self) and substances that are not part of the organism (non-self). The specific immune system includes antibody proteins and other proteins that recognize, bind to, and aid in the destruction of a bacteria or virus. Specific immunity is generally slow in developing but long lasting.

In this investigation, you will develop deeper understanding of first line of defense. You will analyze the role and functions of different mechanisms under the first line of defense.

LESSON 2A ✱ DARE TO INFECT ME

Key Teaching Points

1. An immune system is a network of cells, tissues, and organs that protects animals against pathogens.
2. The defense mechanisms of the immune system can be classified into two main divisions: the innate immune system and the adaptive immune system.
3. Innate immunity is nonspecific immunity and works against a wide variety of invasive organisms.

Learner Outcomes

- Name the system that will fight against bacteria to keep our body healthy.
- List different levels of defense mechanisms within our body to keep us healthy.

Skill Level: Advanced, ages 14-18

Success Indicators

- Participants will describe how skin protects the body from pathogens.

Life Skills

Critical Thinking, Planning and Organizing, Sharing, Nurturing Relationships, Social skills, Communication

Time Needed

- 45 minutes for setting up the learning activity
- Five days for recording results
- 30 minute class discussion and lab write-up

Space

Table and chairs

Suggested Group Size

3 to 4 participants per group

Materials List

per group

- 3 green bananas
- 1 rotten banana
- Spoon (one or more)
- Craft sticks (several)
- Small bowl or plate
- Knife
- Antibacterial soap
- Permanent marker
- 3 sealable plastic bags lettered A-C with a permanent marker
- Paper towels
- Petroleum jelly
- Dare to Infect Me Introduction and Lab handout (1 per participant)

This lesson presents the opportunity to discuss flesh eating bacteria. This link may help answer any questions participants may have: https://www.cdc.gov/features/necrotizingfasciitis/index.html

LESSON 2A DARE TO INFECT ME

Before the Activity

1. Review the Dare to Infect Me presentation.
2. Prepare a projector and screen or other technology to show the presentation to the learners.
3. **This activity needs one rotten banana per group. It takes approximately 2 weeks to rot the banana. Alternatively, you may ask a local grocer to supply you with rotten fruit.**
4. Review the lab instructions and prepare the materials for each group.
5. Print copies of the Dare to Infect Me Introduction and Lab handout.

Follow up for Dare to Infect Me

1. After at least five days have passed, conduct a class wherein participants will complete the Dare to Infect Me lab write-up and class discussion.
2. A projector and screen or other technology is necessary for this class as well.

PART ONE

Opening Question

What happened to the body parts of this animal?

Let's Do It!

1. Show the Dare to Infect Me presentation.
   Ask the Opening Question and facilitate participant discussion.

- What happened to the body parts of this animal?
- Why was the skeleton not disturbed?
- Can bacteria do that when the animal is alive?
- Can they decompose your body? Why not?
- Who or what protects you from bacteria entering your body? (The immune system)

- What is the first thing that protects us from bacteria?
- Explain that any barrier which protects us from the invasion of bacteria is our first line of defense.

2. Give each group of participants the lab materials (if the lab will be conducted in class), along with the learning experience instructions.

3. Each participant will read the introduction to the Dare to Infect Me activity and discuss the pre-lab questions with their groups.

4. Show the Dare to Infect Me video. Make sure to pause the video for prediction. Only show the end of the video if your class will not be conducting the lab in class.

5. Note: If the participants will not be conducting the lab in class, use the video material and demonstration to have participants predict and reflect on the experience shown in the video using the lab worksheets.

6. Participants will follow the lab instructions to complete Dare to Infect Me.

7. Over the next 4 days, participants should discover that the unpeeled banana essentially remains unchanged, as does the banana with the petroleum jelly, while Banana C exhibits changes.

PART TWO

Note: If the participants will not be conducting the lab in class, continue immediately into Part Two and omit the first step.

1. After the fourth day, participants will share their observations with each other and infer that banana peel and petroleum jelly protected the banana against infection, just like our skin and mucus.

2. Ask the Share questions.

   - Which banana was most infected?
   - What was your observation of Banana B?
   - What was your observation of Banana A?

3. Ask the Reflect Questions.

   - Why do you think Banana A did not become infected?
   - Which organ can you relate to the outer covering of a banana?
   - Can you relate any part or product of your body to the petroleum jelly?

- Even though your nose is an entry place for bacteria, why do your lungs not get infected all the time?
- Even though your mouth is an entry place for bacteria, why do you not get sick all the time? What kills those bacteria?
- Even though your eyes are an entry place for bacteria, why don't you get sick all the time? What kills those bacteria?

4. Ask the Generalize question.

   - What is an immune system?

5. Ask the Apply question.

   - Why do cuts on your skin need to be cleaned and bandaged?

Talk It Over

SHARE 1	REFLECT 2
Which banana was most infected? What was your observation of Banana B? What was your observation of Banana A?	Why do you think Banana A did not become infected? Which organ can you relate to the skin of a banana? Can you relate any part or product of your body to the petroleum jelly? Even though your nose is a entry place for bacteria, why do your lungs not get infected all the time? Even though your mouth is an entry place for bacteria, why do you not get sick all the time? What kills those bacteria? Even though your eyes are an entry place for bacteria, why don't you get sick all the time? What kills those bacteria?
GENERALIZE 3 What is an immune system?	
APPLY 4 Why do cuts on your skin need to be cleaned and bandaged?	

LESSON 2A DARE TO INFECT ME

Term and Concept Discovery

Adaptive immunity: distinguishes between substances made by the organism (self) and substances that are not part of the organism (non-self); includes antibody proteins and other proteins that recognize, bind to, and aid in the destruction of a bacteria or virus

First line of defense: part of the innate immune system that includes barriers such as the skin, tears, mucus, and low pH

Immune system: a network of cells, tissues, and organs that protects the human body

Innate immunity: nonspecific immunity that works against wide variety of invasive organisms

Second line of defense: part of the innate immune system that includes phagocytic cells which ingest foreign cells and other particles

LEARN MORE

https://www.cdc.gov/healthyschools/bam/teachers/epi.html

Education Standard(s)

HS-LS1-2, HS-LS1-4 Models (e.g., physical, mathematical, computer models) can be used to simulate systems and interactions—including energy, matter, and information flows—within and between systems at different scales.
HS-LS1-3 Stability and Change: Feedback (negative or positive) can stabilize or destabilize a system.
HS-LS2-8 Empirical evidence is required to differentiate between cause and correlation and make claims about specific causes and effects.

Tags

Immune system, first line of defense, second line of defense, third line of defense

LESSON 2B CODE RED!!

HOW THE HUMAN BODY IS PROTECTED FROM PATHOGENS

Introduction

An **immune system** is a network of cells, tissues, and organs that protects animals against pathogens. The defense mechanisms of the immune system can be classified into two main divisions.

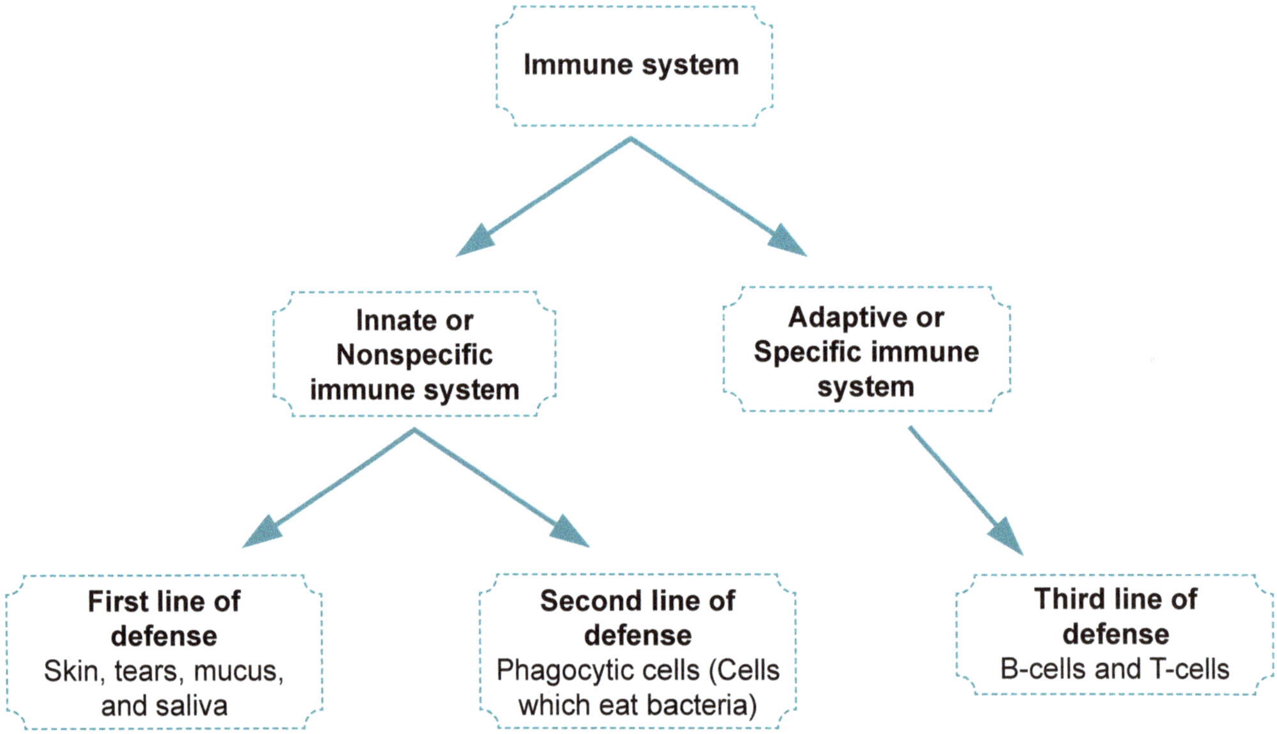

The Innate immune system includes the First line of defense and the Second line of defense. The first line of defense includes barriers such as the skin, tears, mucus, and low pH (acidic) that are toxic to or destroy invaders. The second line of defense includes phagocytic cells that ingest foreign cells and other particles.

Adaptive immunity is **specific**. If a **pathogen** escapes the body's first and second lines of defense, the immune system attacks the pathogen with the **Third line of defense**. The Third line of defense is a **specific immune system. This system fights a specific bacterium with a specific defense**. The specific immune system includes antibody proteins and other proteins that recognize, bind to, and aid in the destruction of bacteria or viruses. Adaptive immunity is generally slow in developing, but long lasting.

Viruses, bacteria, and other pathogens have protein molecules called antigens on their surfaces. When the immune system detects the presence of **antigens** in our body, the immune system triggers the response by producing **antibodies**. An antibody is shaped like the letter "Y" and has two identical antigen-binding sites.

Lesson 2B ✱ Code Red!!

Antibodies are proteins that recognize and bind to antigens and destroy pathogens. These antibodies are carried in the bloodstream to attack the pathogen that is causing the infection.

The specific immune system responds to bacteria, viruses, or pathogens with two types of white blood cells, or **lymphocytes: B lymphocytes (B-cells)** and **T lymphocytes (T-cells)**.

B-cells provide defense (immunity) against antigens (protein molecules on the surface of bacteria). When bacteria invade the body, their **antigens** are recognized by the B-cells. These B-cells grow and divide rapidly, producing large numbers of antibodies. These antibodies seek out the bacteria and lock onto them, causing their destruction.

T-cells provide a defense **against infected cells and abnormal cells. Viruses cannot replicate, or copy, themselves; they must take over a healthy cell to produce more virus.** Cells that are infected by invading viruses are killed by T-cells. When antigens are recognized on an infected cell, T-cells attack and kill the infected cells by creating a hole in the cell and releasing cytotoxins inside the cell, killing all of the virus inside the cell. T-cells also kill abnormal cells like cancerous cells.

In this investigation, you will develop a model of bacteria, B-cells, T- cells, and an infected human cell, and will simulate the immune response to the invasion of bacteria/virus.

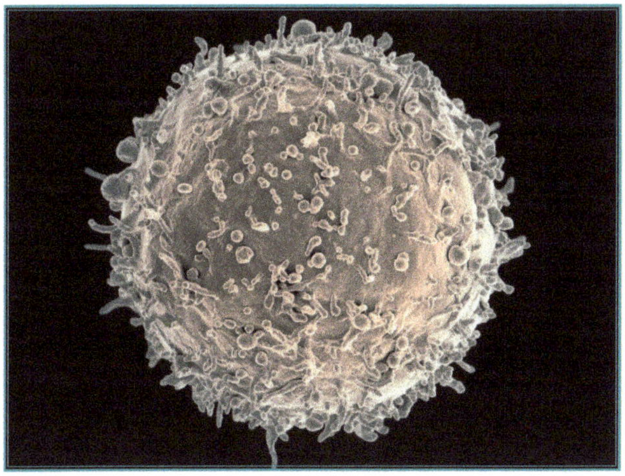

Electron micrograph of a macrophage. Credit: NIAID

Lesson 2B Code Red!!

Key Teaching Points

1. The defense mechanisms of the immune system can be classified into two main divisions: the innate immune system and the adaptive immune system.
2. Adaptive immunity is specific and includes antibody proteins and other proteins that recognize, bind to, and aid in destruction of bacteria or viruses.

Learner Outcomes

- Explore the third line of defense in the immune system.
- Analyze the interaction between antigen and antibody.
- Describe the process by which B-cells and T-cells help our bodies against pathogens.

Skill Level: Advanced, ages 14-18

Success Indicators

- Participants will model the adaptive immune system.
- Participants will describe how the adaptive immune system protects the body from invading bacteria and viruses.

Life Skills

Critical Thinking, Planning and Organizing, Sharing, Nurturing Relationships, Social skills, Communication

Time Needed

- 45-60 minutes for learning experience and write-up

Space

Table and chairs

Suggested Group Size

3 to 4 participants per group

Materials List

For a group of four participants

- 1 braided candy rope (1” to 2” segment per group)
- Gummy bears (minimum 24 per group)
- Round toothpicks (minimum 12 per group)
- Scissors
- Gummy rings (2 per group)
- Juice-filled fruit snacks (6 per group)
- Large round cookies (1 per group)
- Paper towels
- Code Red Introduction and Lab handout (1 per participant)

Please be aware of any food allergies participants may have and take precautions as necessary.

Lesson 2B ✱ Code Red!!

Before the Activity

1. Review the Introduction to Code Red presentation.
2. Prepare a projector and screen or other technology to show the presentation to the participants.
3. Review the lab instructions and prepare the materials for each group ahead of the learning activity.
4. Print copies of the Code Red Introduction and Lab handout.

Code Red!

1. Review the Dare to Infect Me learning experience and the immune system.

2. Show the Introduction to Code Red presentation.

3. Give each group of participants the lab materials, along with the learning experience instructions.

4. Each participant will read the Introduction to the Code Red activity and discuss the pre-lab questions with their groups.

5. Participants will work in groups of four to complete the activity instructions.

Lesson 2B ✹ Code Red!!

Questions and discussions are incorporated into the Lab handout for use in small groups.

Talk It Over

Share 1	Reflect 2
What are the components of the third line of defense? What are antigens? What are antibodies? What produces antibodies? Which two cells respond to bacteria in the specific immune system?	Describe what happens when pathogens such as viruses are able to infect a host cell.
Generalize 3	**Apply 4**
Differentiate between Innate (nonspecific) immunity vs. Adaptive (specific) immunity.	Modeling activity and description of the adaptive immune system during the learning experience.

Term and Concept Discovery

Adaptive immunity: distinguishes between substances made by the organism (self) and substances that are not part of the organism (non-self); includes antibody proteins and other proteins that recognize, bind to, and aid in the destruction of a bacteria or virus

Antibodies: proteins that recognize and bind to antigens and destroy pathogens

Antigens: protein molecules on the surfaces of viruses, bacteria, and other pathogens

First line of defense: part of the innate immune system that includes barriers such as the skin, tears, mucus, and low pH

Immune system: a network of cells, tissues, and organs that protects the human body

Innate immunity: nonspecific immunity that works against wide variety of invasive organisms

Pathogen: a disease-causing agent, such as a microbe, that harms the host organism

Second line of defense: part of the innate immune system that includes phagocytic cells which ingest foreign cells and other particles

Third line of defense: part of the adaptive immune system that includes antibody proteins and other proteins that recognize, bind to, and aid in destruction of bacteria or viruses

Lesson 2B ✹ Code Red!!

Education Standard(s)

HS-LS1-2, HS-LS1-4 Models (e.g., physical, mathematical, computer models) can be used to simulate systems and interactions—including energy, matter, and information flows—within and between systems at different scales.
HS-LS1-3 Stability and Change: Feedback (negative or positive) can stabilize or destabilize a system.
HS-LS2-8 Empirical evidence is required to differentiate between cause and correlation and make claims about specific causes and effects.

Tags

Immune system, first line of defense, second line of defense, third line of defense

Lesson 2A Print Materials

Facilitator Print Materials
None

Group Handouts
None

Youth Handouts
Dare to Infect Me Introduction and Lab handout

See page 3 for access to digital assets & extra enrichment.

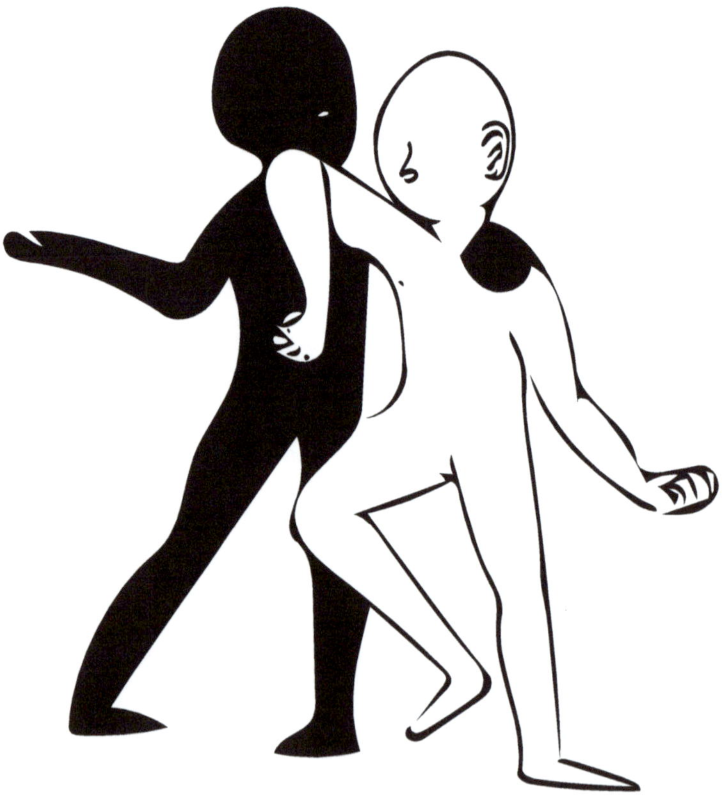

The human body possesses copious defense mechanisms against pathogens and harmful microbial life. This network of cells, tissues, and organs are all a part of the immune system. In this lesson, you will learn two subclasses of the immune system and develop a deeper understanding of the first and second lines of defense.

DARE TO INFECT ME!

Introduction

An immune system is a network of cells, tissues, and organs that protects animals against pathogens. The defense mechanisms of the immune system can be classified into two main divisions.

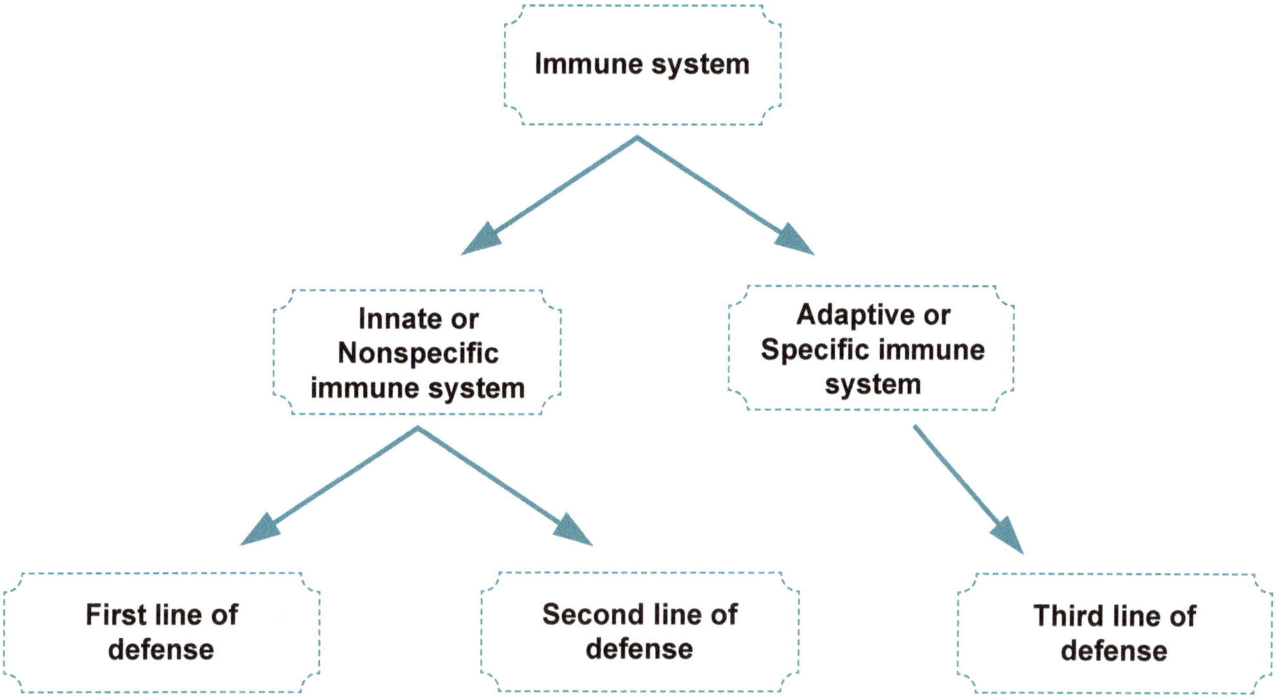

Innate immunity is **nonspecific** immunity and works against a wide variety of invasive organisms. Innate immunity systems include the First line of defense and the Second line of defense. The first line of defense includes barriers such as the skin, tears, mucus, and low pH (acidic) that are toxic to or destroy invaders. The second line of defense includes phagocytic cells that ingest foreign cells and other particles.

Adaptive immunity is **specific** immunity. It distinguishes between substances made by an organism (self) and substances that are not part of an organism (nonself). The specific immune system includes antibody proteins and other proteins that recognize, bind to, and aid in the destruction of bacteria or viruses. It is generally slow in developing, but long lasting.

In this investigation, you will develop a deeper understanding of the first line of defense. You will analyze the role and functions of different mechanisms under the first line of defense.

Pre-lab questions: Discuss the following Pre-lab questions with your group after reading the introduction to the activity.

1. What are the two main divisions of the immune system?
2. What protects us from the millions of bacteria around us?
3. What is the role of the first line of defense mechanisms?

DARE TO INFECT ME! CONTINUED

Materials List

- 3 good bananas
- 1 rotten banana
- Spoon (one or more)
- Craft sticks (several)
- Small bowl or plate
- Knife
- Antibacterial soap
- Permanent marker
- 3 sealable plastic bags lettered A-C with a permanent marker
- Paper towels
- Petroleum jelly
- Lab handout (1 per participant)

Procedure

- Label 3 sealable plastic bags A, B, and C.
- Scoop out a spoonful of rotten banana and keep it in a small bowl or on a small plate.
- Wash your hands and the 3 good bananas with antibacterial soap and dry with a paper towel.
- Keep one banana intact (Banana A).
- Spread a small amount of rotten banana on the surface of Banana A.
- Place Banana A in bag A.
- Carefully push the air out of the bag and seal it completely.
- Cut a 2" strip of peel off of both Banana B and Banana C.
- Spread a generous amount of petroleum jelly on the peeled part of Banana B. Be careful to cover the entire area completely and to spread the petroleum jelly past the edges of the peeled section.
- Spread a small amount of rotten banana on top of the petroleum jelly.
- Place Banana B in bag B.
- Carefully push the air out of the bag and seal it completely.
- Spread a small amount of rotten banana on top of the peeled part of Banana C.
- Place Banana C in bag C.
- Carefully push the air out of the bag and seal it completely.
- Throw the rotten banana away and wash your hands.
- Store the bags for 5 days in a warm, dark place.
- Observe the bananas daily for five days. DO NOT open the bags. Record your observations and return the bags to their storage place each day.

DARE TO INFECT ME! OBSERVATIONS

Pre observation questions:

What do you predict for each banana? Why?

Post observation questions:

Was your prediction correct for each banana?
How did the appearance of each banana change in 5 days?
What caused the change (if any) or no change in the appearance of the bananas?
How do the bananas represent real life situations? Be sure to include the role of the rotting banana in your discussion.
Explain why cuts on your skin should be cleaned and bandaged.
Even though your nose is an entry place for bacteria, why do your lungs not get infected all the time?

DARE TO INFECT ME! OBSERVATIONS CONTINUED

Even though your mouth is an entry place for bacteria, why do you not get sick all the time? What kills those bacteria?
Even though your eyes are an entry place for bacteria, why do you not get sick all the time? What kills those bacteria?

Observation Table:

Days	Banana A	Banana B	Banana C
One			
Two			
Three			
Four			
Five			

Lesson 2B Print Materials

Facilitator Print Materials
None

Group Handouts
None

Youth Handouts
Code Red Introduction and Lab handout

In the previous lesson, the immune system was divided into two categories: the innate and adaptive systems. The adaptive system is specific and reactive. In this lesson, you will explore the third line of defense and interactions between antigens and antibodies within the immune system.

CODE RED!! HERE COMES THE THIRD LINE OF DEFENSE

Recall that an immune system is a network of cells, tissues, and organs that protects animals against pathogens. The defense mechanisms of the immune system can be classified into two main divisions.

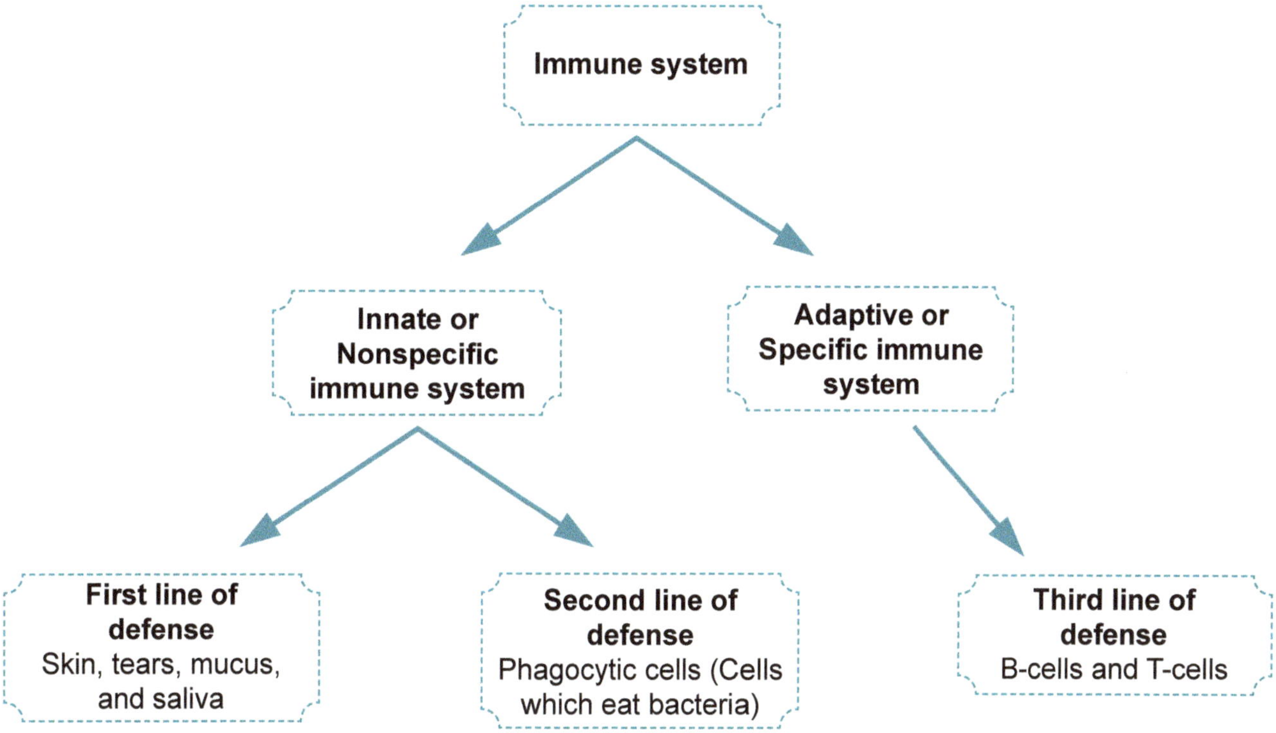

In Dare to Infect Me, you learned about the Innate immunity system, which includes the First line of defense and the Second line of defense. Remember, the first line of defense includes barriers such as the skin, tears, mucus, and low pH (acidic) that are toxic to or destroy invaders. The second line of defense includes phagocytic cells that ingest foreign cells and other particles.

Adaptive immunity is **specific**. If a pathogen escapes the body's first and second lines of defense, the immune system attacks the pathogen with the **Third line of defense**. The Third line of defense is a **specific immune system. This system fights a specific bacterium with a specific defense.** The specific immune system includes antibody proteins and other proteins that recognize, bind to, and aid in the destruction of bacteria or viruses. Adaptive immunity is generally slow in developing, but long lasting.

Viruses, bacteria, and other pathogens have protein molecules called antigens on their surfaces. When the immune system detects the presence of **antigens** in our body, the immune system triggers the response by producing **antibodies**. An antibody is shaped like the letter "Y" and has two identical antigen-binding sites. Antibodies are proteins that recognize and bind to antigens and destroy pathogens. These antibodies are carried in the bloodstream to attack the pathogen that is causing the infection.

The **specific immune system** responds to bacteria, viruses, or pathogens with two types of white blood cells, or **lymphocytes: B lymphocytes (B-cells)** and **T lymphocytes (T-cells)**.

Code Red!! Here Comes the Third Line of Defense ...Continued

B-cells provide a defense (immunity) against antigens (protein molecules on the surface of bacteria). When bacteria invade the body, their **antigens** are recognized by the B-cells. These B-cells grow and divide rapidly, producing large numbers of antibodies. These antibodies seek out the bacteria and lock onto them, causing their destruction.

T-cells provide a defense **against infected cells and abnormal cells. Viruses cannot replicate, or copy, themselves; they must take over a healthy cell to produce more virus**. Cells that are infected by invading viruses are killed by T-cells. When antigens are recognized on an infected cell, T-cells attack and kill the infected cells by creating a hole in the cell and releasing cytotoxins inside the cell, killing all of the virus inside the cell. T-cells also kill abnormal cells like cancerous cells.

In this investigation, you will develop models of bacteria, B-cells, T- cells, and an infected human cell, and will simulate the immune response to the invasion of bacteria/virus.

Pre-lab questions: Discuss the following Pre-lab questions with your group after reading the introduction to the activity.

1. Name the system that protects against pathogens.
2. Differentiate between Innate (nonspecific) immunity vs. Adaptive (specific) immunity.
3. What are two types of the body's nonspecific defense mechanisms?
4. What is the role of skin, tears, mucus, and saliva in the first line of defense?
5. What is the role of phagocytic cells in the Second line of defense?
6. What are the components of the Third line of defense?
7. What are antigens?
8. What are antibodies?
9. What produces antibodies?
10. Which two cells respond to bacteria in the specific immune system?
11. Describe what happens when pathogens such as viruses are able to infect a host cell.

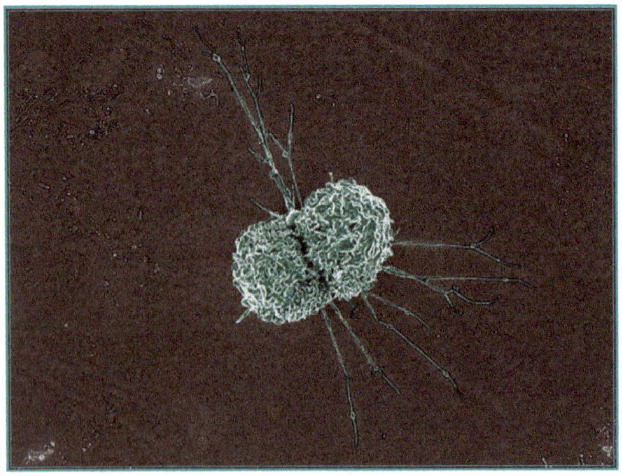

Electron micrograph of a B cell. Credit: NIAID

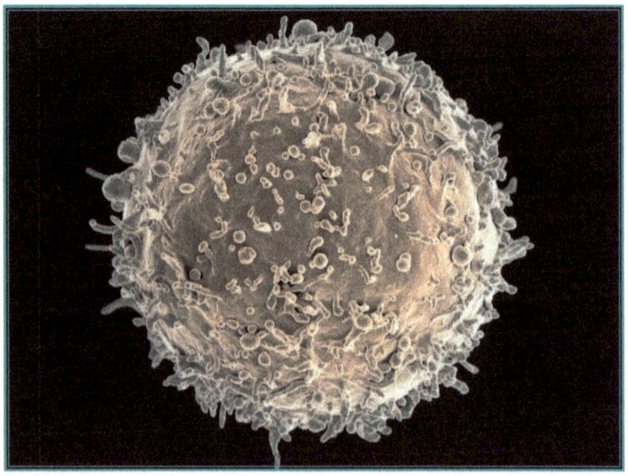

Electron micrograph of a macrophage. Credit: NIAID

Code Red!! Lab

Material List:

- 1 braided candy rope (1" to 2" segment per group)
- Gummy bears (minimum 24 per group)
- Round toothpicks (minimum 12 per group)
- Scissors
- Gummy rings (2 per group)
- Juice-filled fruit snacks (6 per group)
- Large round cookies (1 per group)

Procedure:

Step 1 Build your Bacteria:

- A segment of candy rope will represent your bacteria. Bacteria have antigens on their surface. (An antigen is a substance that stimulates your immune response to produce antibodies.)
- Using scissors, cut 3 toothpicks in half.
- Using scissors, cut 6 gummy bears into triangular prisms.
- Insert the pointed ends of each half toothpick into the triangular gummies (antigens). Then, stick the 6 antigens into the bacteria (candy rope).
- Your bacteria is ready!

Step 2 Build your B-cell:

- A gummy ring represents your B-cell. B-cells produce antibodies. (Antibodies are protein molecules, produced by B-cells, which bind to antigens on bacteria or any pathogen and immobilize that pathogen.)
- Using scissors, cut 3 toothpicks in half.
- Using scissors, cut 6 gummy bears into a 'Y' shape (antibodies).
- Insert the pointed ends of each half toothpick into one of the cut gummy bears (these are antibodies).
- Stick the 6 antibodies into the B-cell (gummy rings).
- Your B-cell is ready!

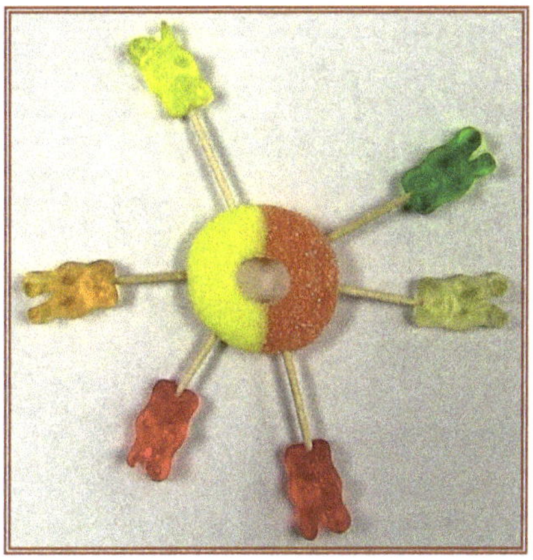

CODE RED!! LAB CONTINUED

Step 3 Build your infected cell:

- A large cookie represents your infected cell. Infected cells have antigens on
- their surfaces and have been infected by a
- virus, which cannot replicate (copy) itself.
- Using scissors, cut 3 toothpicks in half.
- Using scissors, cut 6 gummy bears into triangular prisms.
- Insert the pointed ends of each half toothpick into the triangular gummies (antigens). Then, stick the 6 antigens into the infected cell (large cookie).
- Your infected cell is ready!

Step 4: Build your T-cell:

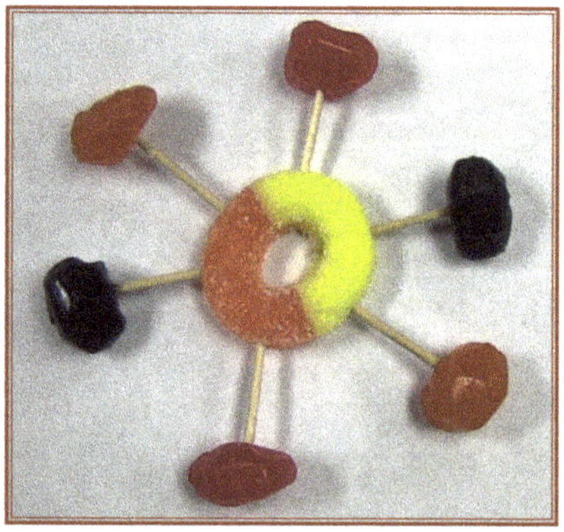

- A gummy ring represents your T-cell. T-cells have receptors on their surfaces. (Receptors attach to infected cell's antigens, which allows the T-cell to pump cytotoxins into the infected cell, thus effectively killing the infected cell.)
- Using scissors, cut 3 toothpicks in half.
- Insert the pointed ends of each half toothpick into a juice-filled candy (receptor). Then, stick the 6 receptors into the gummy ring. (T-cell). (Be careful to not spill all the juice onto your workstation, it is advised you stick the toothpicks in over the top of a paper towel.)
- Your T-cell is ready!

Step 4 Simulate:
Model the process of B-cells defending against the pathogen and orally explain the process to your group members until all are able to explain the process.

Step 5 Simulate:
Model the process of the T-cells defending against the infected cell and orally explain the process to your group members until all are able to explain the process.

Lesson 3 Outbreak!

Investigating the Epidemiology of Bacterial Infections

Introduction

This is a Web-based research activity in which participants are introduced to the world of epidemiology through the Centers for Disease Control CDC website (http://www.cdc.gov/az/). Participants research major world diseases to learn about the detection of an outbreak, the cause of an outbreak, the measure of an outbreak, and epidemiologic inference and action plan taken for the prevention of the specific disease.

This activity will be completed in groups of two participants. Each group will research their assigned disease using the CDC website and then create a display using the Outbreak Trifold Template.

Participants will explain and discuss various tools used by epidemiologists to investigate, analyze, measure, and design an action plan to prevent the outbreak of the disease.

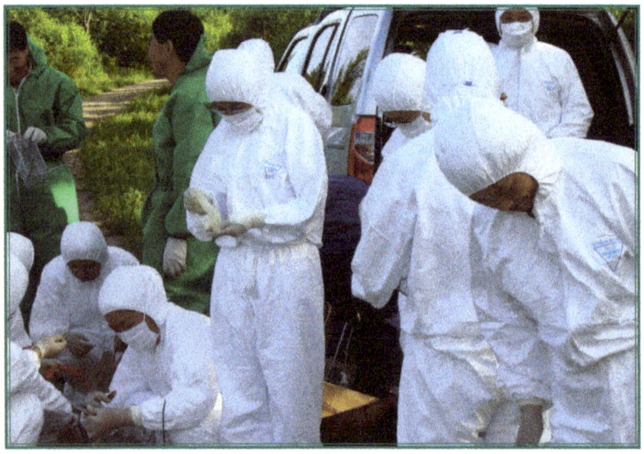

Chinese epidemiologists, in the process of collecting data. (https://www.cdc.gov)

Field Epidemiology Training Program (FETP) resident in the laboratory. (https://www.cdc.gov)

The Epidemic Intelligence Service (EIS) officer conducting an immunization coverage survey. (https://www.cdc.gov)

Lesson 3 Outbreak!

Key Teaching Points

1. Infectious diseases can be spread from person to person or through exposure to infectious agents in our environments.
2. Outbreaks of infectious disease vary and are controlled by methods specific to the disease.

Learner Outcomes

- Find, summarize, and explain information about an infectious disease.
- Construct a museum display by organizing the information about an infectious disease.
- Compare the outbreak, period of outbreak, and control of outbreak for these diseases.

Skill Level: Advanced, ages 14-18

Success Indicators

- Participants will research an outbreak on the CDC website or other websites or print materials provided by the facilitator.
- Participants will produce an informative product presenting the progress of a disease outbreak.

Life Skills

Sharing, Nurturing Relationships, Social skills, Communication, Keeping Records, Teamwork

Time Needed

Four sessions of 90 Minutes each.

Space

Workspace with internet access

Suggested Group Size

2 participants per group

Materials List

- Web enabled computers
- Outbreak presentation
- 1 prepared notecard per participant
- 1 copy of the article per participant
- 1 Outbreak Trifold Template per pair of participants
- Tri fold poster board

Websites that provide information about diseases, including the CDC, can contain graphic and disturbing images. Please preview the websites that participants will use or provide printed information for their use.

Lesson 3 ✱ Outbreak!

Before the Activity

1. Make note cards, 1 per participant, with the name of a hotel on one side. Three cards should be labeled Holiday Inn. The other cards should have names of other local hotels, approximately three cards per hotel.
2. Prepare copies of the News Article: "Salmonella Outbreak Linked to N.C. Holiday Inn Sickened 100, Final Report Says."
3. Prepare copies of the Outbreak Trifold Template, 1 per pair of participants.
4. Set up technology in the classroom for participant research.
5. Become familiar with the CDC website (http://www.cdc.gov/az/). Diseases for participants to research may be chosen ahead of time, or participants may be allowed to pick diseases for themselves.
6. Set up an area of the classroom space with materials for completing the project: trifold boards, project supplies as necessary.
7. Prepare technology to show the following video. https://youtu.be/UG8YbNbdaco

Opening Questions

Can it happen to me?

Let's Do It!

1. Give each participant a notecard with a name of a hotel on one side.

2. Describe the scenario to the class: They all are here for a conference and staying at different hotels. Today is the first day they are meeting one another and getting to know each other. While they are getting to know each other, each person will meet five people in the room and request for them to write their name on one side of the notecard.

3. Distribute the News Article: **"Salmonella Outbreak Linked to N.C. Holiday Inn Sickened 100, Final Report Says."**

4. Ask everyone, Who is staying at the Holiday Inn? How many people came in contact with the infected person? Anyone who came in contact with anyone staying at the Holiday Inn is infected by Salmonella as well. If anyone had contact with an infected person they are also infected.

5. Final result: the whole class is infected.

6. Show the following video to the participants. https://youtu.be/UG8YbNbdaco

7. Assign each pair of participants a disease or allow them time to choose one from the CDC website.

8. Each group shall research their disease using the CDC website and then create a display on the trifold poster using words and pictures to answer the questions on the Outbreak Trifold Template.

9. Participants will present their projects to the class to share and discuss various tools used by epidemiologists to investigate, analyze, measure, and design the action plan to prevent the outbreak of the disease. **Option:** Participants may present projects to a larger audience, either in the community or within the learning environment.

10. Ask the Generalize question and help participants to compare and contrast the findings of their research.

    - Based on your historical research on outbreaks, how were outbreaks prevented and managed?

11. Ask the Apply question and help participants to extend the information they have learned into their environment.

    - How will you make sure you and your community are not part of an outbreak of infectious disease?

Talk It Over

Share 1	Reflect 2
Participants will present their projects to the class to share and discuss various tools used by epidemiologists to investigate, analyze, measure, and design the action plan to prevent the outbreak of the disease.	Facilitator will discuss the Analysis questions of the investigation with the help of the presentation: Why are some diseases concentrated in specific geographical area?
Generalize 3	**Apply 4**
Based on your historical research on outbreaks, how were outbreaks prevented and managed?	Participants will present their projects to the class to share and discuss various tools used by epidemiologists to investigate, analyze, measure, and design the action plan to prevent the outbreak of the disease.

Lesson 3 ✹ Outbreak!

Term and Concept Discovery

Epidemiology: the branch of medicine dealing with the incidence and prevalence of disease in large populations and with detection of the source and cause of epidemics of infectious disease.

Outbreak: a sudden, violent, or spontaneous occurrence, esp of disease or strife.

Preventive measures: a variety of interventions that can be undertaken to prevent or delay the occurrence of disease or reduce further transmission or exposure to disease

Vaccinations: Inoculation with a vaccine in order to protect against a particular disease.

References

http://www.cdc.gov/az/

Variations

Instead of creating a trifold display, participants can use a different platform to share their research, such as a PowerPoint or video.

Education Standard(s)

Bio.2.1.4 Explain how disease can disrupt ecosystem balance. (Examples: AIDS, influenza, tuberculosis, Dutch elm disease, Pfiesteria, etc.)
Bio.2.2.1 Infer how human activities (including population growth, pollution, global warming, burning of fossil fuels, habitat destruction and introduction of nonnative species) may impact the environment.
Bio.3.4.3 Explain how various disease agents (bacteria, viruses, chemicals) can influence natural selection.
HS-LS1-3 Plan and conduct an investigation individually and collaboratively to produce data to serve as the basis for evidence, and in the design: decide on types, how much, and accuracy of data needed to produce reliable measurements and consider limitations on the precision of the data (e.g., number of trials, cost, risk, time), and refine the design accordingly.
HS-LS1-3 Scientific inquiry is characterized by a common set of values that include: logical thinking, precision, open-mindedness, objectivity, skepticism, replicability of results, and honest and ethical reporting of findings.
HS-LS2-2, HS-LS2 A complex set of interactions within an ecosystem can keep its numbers and types of organisms relatively constant over long periods of time under stable conditions. If a modest biological or physical disturbance to an ecosystem occurs, it may return to its more or less original status (i.e., the ecosystem is resilient), as opposed to becoming a very different ecosystem. Extreme fluctuations in conditions or the size of any population, however, can challenge the functioning of the ecosystems in terms of resources and habitat availability.

Tags

Epidemiology, Outbreak, CDC, Prevention, Data and Statistics

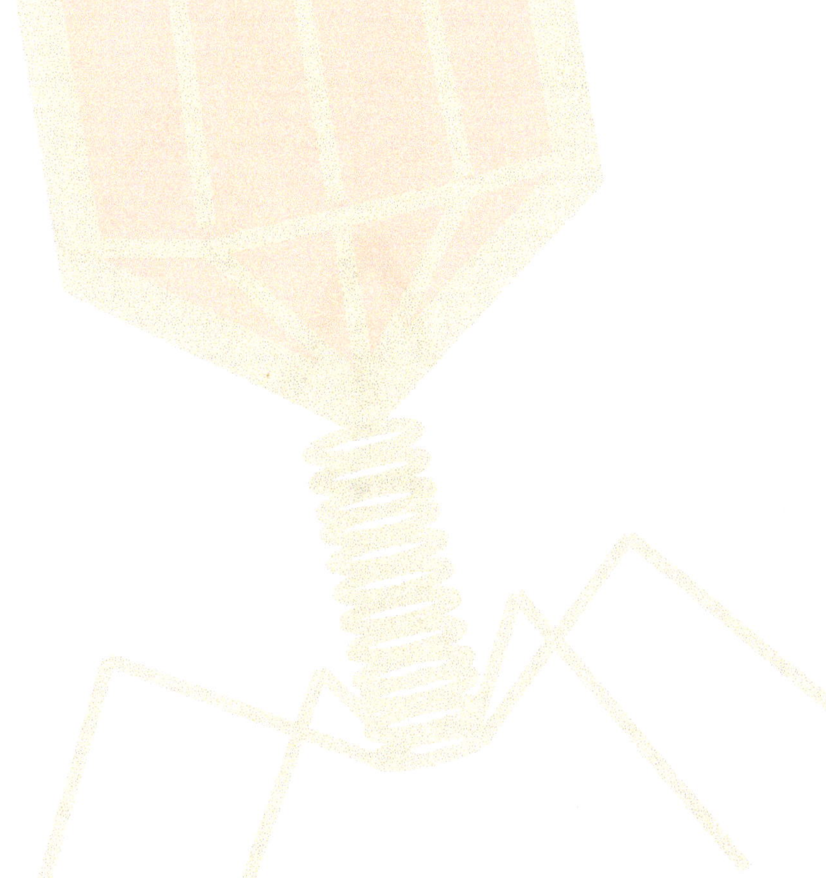

Lesson 3 Print Materials

Facilitator Print Materials

None

Group Handouts

Holiday Inn Article

Youth Handouts

Outbreak Trifold Template

See page 3 for access to digital assets & extra enrichment.

Pathogens can be contagious and sporadically spread from person to person. Epidemics within a population can be controlled with containment protocols specific to the disease. In this lesson, you will learn about infectious diseases and the outbreak, period of outbreak, and control of outbreak for these diseases.

Food Safety News

Breaking news for everyone's consumption

Home | Foodborne Illness Outbreaks | Food Recalls | Food Politics | Events | Subscribe | About Us

Salmonella Outbreak Linked to N.C. Holiday Inn Sickened 100, Final Report Says

BY GRETCHEN GOETZ | JULY 23, 2013

The Salmonella Typhimurium outbreak linked to a Holiday Inn in North Carolina ultimately sickened at least 100 people this May, the North Carolina Division of Public Heath reported this week.

The bacteria was linked to the All American Grill at the Holiday Inn Bordeaux in Fayetteville, NC, according to the final outbreak report, issued by NCDPH July 19.

Health officials in Cumberland County first alerted the state health department of a Salmonella outbreak that had sickened an estimated 15 people on May 15. All of the ill people had reported eating at the Holiday Inn Bordeaux in the days preceding illness. The case count would eventually expand to 88 – the number reported May 28. Now, in its final outbreak report, NCDPH says 100 Salmonella illnesses were ultimately linked to the outbreak.

That final count included 25 cases confirmed by laboratory testing and an additional 75 suspected cases. Of those sickened, 29 were staff members at the hotel. Victims ranged in age from 17 to 81 years old.

The first illnesses began May 1, 2013 and the last reported illness onset was May 17, 2013.

The strain of Salmonella Typhumurium responsible for the outbreak was one that's unique to North Carolina, according to the report.

While the specific vehicle of the bacteria was not identified, health officials found a series of sanitation violations that might have led to contamination.

"Interviews with managerial staff and observation of food preparation identified multiple opportunities for Salmonella contamination, including improper water temperatures and the absence of hand washing supplies in some areas," the final report notes. "Other potential food safety issues that were identified included bare hand contact with ready to eat foods, temperature violations, and a dishwasher in one kitchen that was not operating effectively as described by staff members."

Health officials speculate that various foods and surfaces in the restaurant kitchen were cross-contaminated, which would explain why they were unable to identify the specific source of the bacteria.

General information for the Pathogen

Information of illness caused by the Pathogen: Include symptoms, diagnosis, treatments, and prevention

Title: Name of the Pathogen

Labeled diagram of the Pathogen

Data about the outbreak and graphical analysis

Current measures taken by CDC to control the outbreak

What should be done at school level to increase the awareness about this Pathogen

Acknowledgements

Lesson 4 Knock, Knock – Who's There?

An Introduction to the Tools of Molecular Biology

Introduction

Even though large numbers of bacteria inhabit the human microbiome, it is very difficult to observe these bacteria through growth on a nutrient medium (on a petri dish). In fact, only 1-2% of bacteria can be observed in this manner. One way to identify more of the members of a microbiome is to use DNA sequencing. Distinct DNA sequences can be **analyzed** and compared to existing sequences in a database to identify the number and variety of bacterial species present.

DNA can be **extracted** from any cell, including your cells. It is a simple process involving breaking open the cells with detergent, dissolving the DNA in a salt solution, and then separating the DNA molecules using cold alcohol.

DNA fragments can be **separated** by placing samples in a gel matrix and applying a low voltage current. Because DNA is negatively charged, the fragments move through the gel and are distributed according to size – smaller molecules travel faster and larger molecules are retained by the gel matrix.

DNA can be **sequenced** by several means, all based on the principle of base pairing – in DNA the base A always pairs with the base T, and G pairs with C. The **Sanger sequencing method** employs building a complementary strand in which the new chain terminates at a given base. By arranging the new sequence in order of increasing length, the order of bases can be determined. A second method called **next-generation sequencing** allows multiple unique DNA fragments to be sequenced simultaneously, greatly increasing the speed with which DNA samples may be identified.

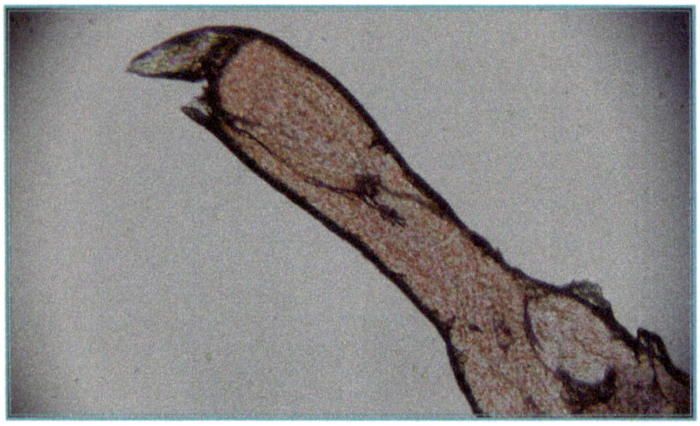

Tomato DNA

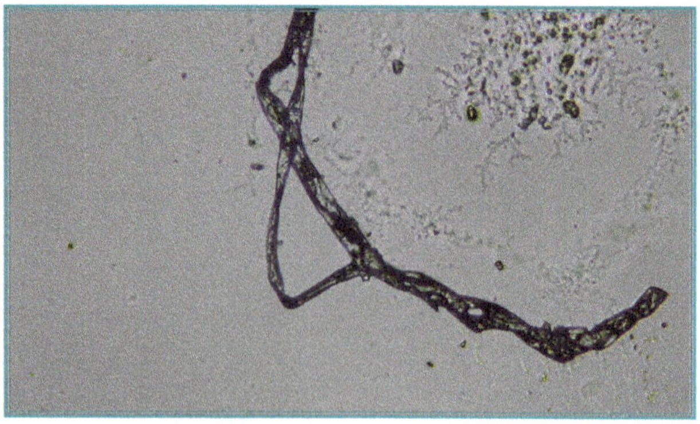

Strawberry DNA

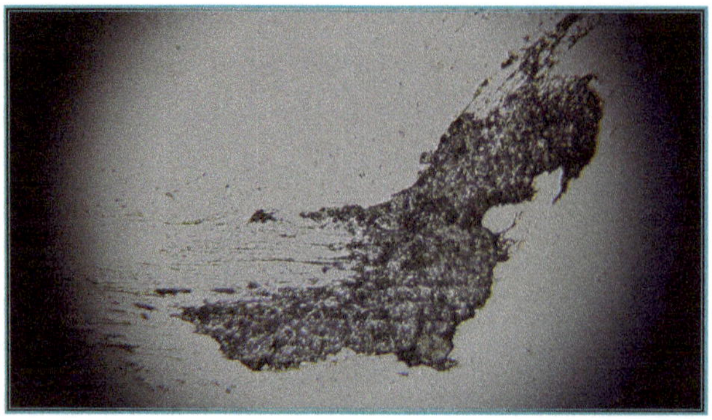

Kiwi DNA

Lesson 4 Knock, Knock - Who's There?

Key Teaching Points

1. DNA can be extracted from any cell.
2. DNA can be sequenced by several means, all based on the principle of base pairing.
3. DNA sequencing methods can be used to identify members of a microbiome.

Learner Outcomes

- Recognize that DNA sequences allow identification and organization of groups of organisms.
- Demonstrate the presence of DNA in cells through DNA extraction.
- Model separation of DNA fragments by size and charge.
- Model DNA sequencing strategies.

Skill Level: Advanced, ages 14-18

Success Indicators

- Participants will identify a "foreign" (non-matching) DNA sequence from a population of representative bacterial genomes.
- Participants will isolate visible DNA from human cells.
- Participants will predict size distribution of DNA fragments from virtual gel electrophoresis separation.
- Participants will build unknown DNA sequences using two methods.

Life Skills

Critical Thinking, Problem Solving, Teamwork

Time Needed

2 days/meetings

Activity 1:

Part A: 20 minutes for "Finding the Culprit"
Part B: 30 minutes for "DNA extraction"

Activity 2:

Part A: 20 minutes for "Sanger Sequencing"
Part B: 30 minutes for "Next Generation Sequencing"

Space

Large meeting room

Suggested Group Size

12-16 in groups of 4

Materials List

Activity 1, Part A:

(For 3 groups)
Finding the culprit in your chicken

- 33 pipecleaners
- colored beads (4 colors)
- 3 ziplock bags (large)

Activity 1 Part B: How is the DNA obtained?

For each participant:

- Bathroom cup (3-4 oz.)
- Test tube
- A pair of safety gloves
- Dropper
- Tall glass or test tube holder

For each group:

- Clear dish detergent
- Sports drink (eg Gatorade)
- Meat tenderizer (eg Adolph's)
- 90% Isopropyl alcohol (rubbing alcohol) - COLD - keep in the freezer until just before use

Activity 2, Part A:

Sanger Sequencing paper modeling

- Sanger Sequencing Simulation instructions (1 per group)
- DNA Sequencing by Chain Termination handout (1 per group)
- Tape
- Scissors (1 per group)

Extension: Activity 2, Part B:

How is the DNA used to identify organisms?

- Next Generation Sequencing Game – see instructions and materials
- "Unknown" DNA sequences – as many as desired. 1 unique sequence per "bead"
- Nucleotide labels – "A" "T" "C" "G" – can have multiple sets
- "Bar code" sequences that match "unknown"
- Chart paper or whiteboard for Reporters
- Markers

Lesson 4 Knock, Knock - Who's There?

Before the Activity

- Review the Presentation for this lesson.

Activity 1, Part A: Finding the culprit in your chicken

Construct gut microbiome "DNA sequences" from pipe cleaners. Thread colored beads onto pipe cleaners. The following is for 3 groups of participants:

- Make 8 pipe cleaners with the same bead sequences (A); make 8 pipe cleaners with a different sequence (B), make 8 pipe cleaners with yet another different sequence (C), and a final 7 pipe cleaners with a fourth different sequence (D). Make 2 pipe cleaners (E) with a sequence that is different from sets A-D. This yields a total of 33 pipe cleaners.
- Place 2 or 3 copies of each pipe cleaner A-D into each of three-gallon zip bags and label the zip bags "Chicken 1," Chicken 2," and "Chicken 3." Place the two copies of sequence (E) into the bag labeled "Chicken 3." A total of 11 pipe cleaners should be in each bag. Participants should identify arrangement "E" as the DNA sequence unique to the chicken that is carrying Salmonella in its gut (Chicken 3).
- Once the sets are constructed they can be used repeatedly. See sample arrangements below: (These would be in the bag labeled "Chicken 3")

DNA strands will need to be created prior to the lesson and may be time intensive. This non-consumable resource can be used for future classes.

Activity 1, Part B: How is the DNA obtained?

- Assemble the materials for DNA extraction.

Activity 2, Part A: How is the DNA used to identify organisms?

- Make copies of the Sanger Sequencing Simulation instructions and the DNA Sequencing by Chain Termination handout for each group.

Lesson 4 Knock, Knock - Who's There?

Extension: Activity 2, Part B: How can multiple different DNA molecules be sequenced at the same time?

- Cut out all print materials for "Next Generation Sequencing Game."

Opening Questions

How can DNA help identify the microbes in a population?

You want to use only chicken that doesn't harbor dangerous bacteria. How can you tell?

Let's Do It!

Activity 1, Part A: Finding the culprit in your chicken

1. Give each group one "Chicken" to analyze.
2. Ask the Opening Questions.
3. Participants will look at the collection of pipe cleaner "DNA Sequences" in each zip bag. These models represent DNA from all of the stomach microorganisms isolated from three chickens: 1, 2, and 3. They only want to raise chickens without the dangerous bacteria in their gut. The DNA was isolated from stool samples because large amounts of excreted material consist of bacteria that are present in the digestive system. Don't worry – it's only DNA now!
4. Participants will sort the set of 11 bacterial sequences from each chicken into groups based on the sequence similarity. Each colored bead represents one of the four bases in DNA – A, C, T, or G. There will be several "bacteria" whose sequences are almost exactly alike – a difference in a base or two is OK.
5. Once the groups finish analyzing their "Chicken," groups will trade bags and analyze the next set of DNA sequences. Continue until all groups have analyzed the information in all three bags.
6. Guide the groups to COMPARE the different bacterial DNA sequences from each chicken. How many different patterns of bacterial DNA were found in each chicken here? Are all three of the chickens' samples the same or very similar? Is there anything different about the bacterial DNA that came from one of the chickens?

7. Ask the Share questions.

    - Were you able to identify the chicken that harbored the dangerous bacteria? How did you do it?

8. Ask the Reflect question.

    - Based on what you learned from Lesson 3, how would a chicken's body try to fight the bacterial infection?

9. Ask the Generalize questions.

    - Can animals have microorganisms in their digestive system and not be sick? How do you know? Refer to the information from Lesson 1 for help.

Activity 1, Part B: How is the DNA obtained?

1. **DNA extraction – from YOU.** With very basic materials, you can see your DNA!

2. "Extracting DNA from Your Cells" handout – use your cheek cells as a source.

3. Ask the Reflect questions.

    - How useful would it be to have the DNA sequence information available for investigating disease outbreaks?
    - How would a person or their doctor conduct the tests and access the information?

(This is a good "Break" for splitting up the activities.)

Activity 2, Part A: How is the DNA used to identify organisms?

1. If a new class period has begun, review the background information on DNA structure, especially the rules for base-pairing. The structure of the DNA molecule is what makes both of these approaches possible and reliable.

2. Show the following video to introduce the process of Sanger sequencing.

    http://www.dnalc.org/view/15479-Sanger-method-of-DNA-sequencing-3D-animation-with-narration.html

3. Use the Paper modeling of Sanger Sequencing activity to find the DNA sequence of a sample of mystery DNA.

Extension: Activity 2, Part B: How can multiple different DNA molecules be sequenced at the same time?

***Note: This learning experience is for advanced classes of participants.**

1. Participants will access and follow the animation of massively parallel sequencing at http://www.yourgenome.org/video/sequencing-at-speed

2. Optional: Review principles of next-generation sequencing in background presentation with participants.

3. Play **Next Generation Sequencing Role-Play game** and sequence several samples in record time! Compare the sequences that are obtained with the "DNA Sequence Database." Does your chicken harbor a dangerous bacterium?

4. Ask the Share questions.

    - After you completed the "Sanger Sequencing" activity, how accurate were your unknown DNA sequences when you compared them to the correct ones? Where did or might errors occur?

    - After you completed the "Next Generation Sequencing" activity, were you able to identify the different bacteria based on their DNA sequence? How accurate was your sequencing?

5. Ask the Reflect question.

    - Would it be troublesome to have your microbiome sequenced and on record somewhere?

6. Ask the Generalize question.

    - Can you think of some other instances where it would be an advantage to sequence a large number of DNA samples as quickly as possible?

7. Ask the Apply questions.

    - Find out about careers that use DNA sequencing capability (for example, forensics, medical treatment, or epidemiology). What kind of training might you need to work in a field like forensics, medicine, or epidemiology?

Lesson 4 Knock, Knock - Who's There?

Talk It Over

Share 1	Reflect 2
Were you able to identify the chicken that harbored the dangerous bacteria? How did you do it? After you completed the "Sanger Sequencing" activity, how accurate were your unknown DNA sequences when you compared them to the correct ones? Where did or might errors occur? After you completed the "Next Generation Sequencing" activity, were you able to identify the different bacteria based on their DNA sequence? How accurate was your sequencing?	Based on what you learned from Lesson 2, how would a chicken's body try to fight the bacterial infection? How useful would it be to have the DNA sequence information available for investigating disease outbreaks? How would a person or their doctor conduct the tests and access the information? Would it be troublesome to have your microbiome sequenced and on record somewhere?
Generalize 3	**Apply 4**
Can animals have microorganisms in their digestive system and not be sick? How do you know? Refer to the information from Lesson 1 for help. Can you think of some other instances where it would be an advantage to sequence a large number of DNA samples as quickly as possible?	Find out about careers that use DNA sequencing capability (for example, forensics, medical treatment, or epidemiology). What kind of training might you need to work in a field like forensics, medicine, or epidemiology?

Term and Concept Discovery

DNA pairing: bases on one strand are paired specifically with a base on the opposite strand – A always pairs with T and C with G

DNA structure: is a double strand with a sequence of four bases along the molecule. Each base on one strand is paired specifically with a base on the opposite strand – A always pairs with T and C with G

- Information is contained in the order of bases along the DNA molecule
- Molecular tools can be used to identify the sequence of DNA molecules
- DNA sequences are unique to individual organisms and can be used for identification

DNA sequences: each DNA sequence is unique to an individual organism and can be used for identification

Lesson 4 ✱ Knock, Knock - Who's There?

Variations

Extract DNA on day/meeting 1
Analyze DNA Sequences on Day/meeting 2 (DNA Sequencing strategies)
Apply procedures on Day/meeting 3 (Find the culprit)

Acknowledgements

"*Extracting Cheek cell DNA*" About.com Biology
"*Bacterial Identification Virtual Lab*" – HHMI
"*Next Generation Role Play*" adapted from principle of Ion Torrent Sequencing from Life Technologies

References

Cheek cell DNA extraction:
http://learn.genetics.utah.edu/content/labs/extraction/howto/
Massively parallel sequencing: http://www.yourgenome.org/video/sequencing-at-speed

Learn More

For a step-by-step virtual investigation of how DNA sequences can be used to identify bacteria, visit the Bacterial Identification Virtual Lab
http://media.hhmi.org/biointeractive/vlabs/bacterial_id/index.html

Grab and GO: DNA separation by gel electrophoresis
You can't just pick up DNA pieces with tweezers! Learn about the basic process for separating different pieces of DNA from a mixture by passing them through a jello-like material using electric current.

Education Standard(s)

NGSS HS-LS3.A Each chromosome consists of a single very long DNA molecule, and each gene on the chromosome is a particular segment of that DNA. The instructions for forming species' characteristics are carried in DNA. All cells in an organism have the same genetic content, but the genes used (expressed) by the cell may be involved in regulatory or structural functions, and some have no as-yet known function
NGSS HS-LS4.A Genetic information provides evidence of evolution. DNA sequences vary among species, but there are many overlaps; in fact, the ongoing branching that produces multiple lines of descent can be inferred by comparing the DNA sequences of different organisms. Such information is also derivable from the similarities and differences in amino acid sequences and from anatomical and embryological evidence.

Tags

DNA structure, DNA sequencing, biotechnology, microbiome

Lesson 4 Print Materials

Facilitator Print Materials
Sanger Sequencing Simulation answer key
Next Generation Sequencing Role-play Game
Barcodes for Reporters
Sequences for Beads
Nucleotide Labels
"Database" of Bacterial DNA Sequences
NGSS Diagram

Group Handouts
Extracting DNA from Your Cells
DNA Sequencing by Chain Termination
Sanger Sequencing Simulation Instructions

Youth Handouts
None

See page 3 for access to digital assets & extra enrichment.

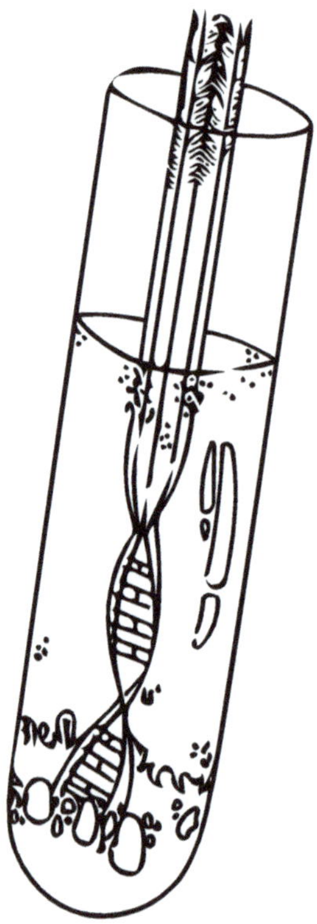

DNA can be extracted from any celled organism. This includes prokaryotic and eukaryotic life. Sequences of DNA can be analyzed by comparison to those in a pre-existing database and used to identify and organize groups of organisms. In this lesson, you will learn about DNA and the extraction, separation, and sequencing methods of DNA.

SANGER SEQUENCING SIMULATION ANSWER KEY

A G G C T C T G A A G T G C T A

Today, automated sequencers "read" the color assigned to each base and record them in order as shown in the example below:

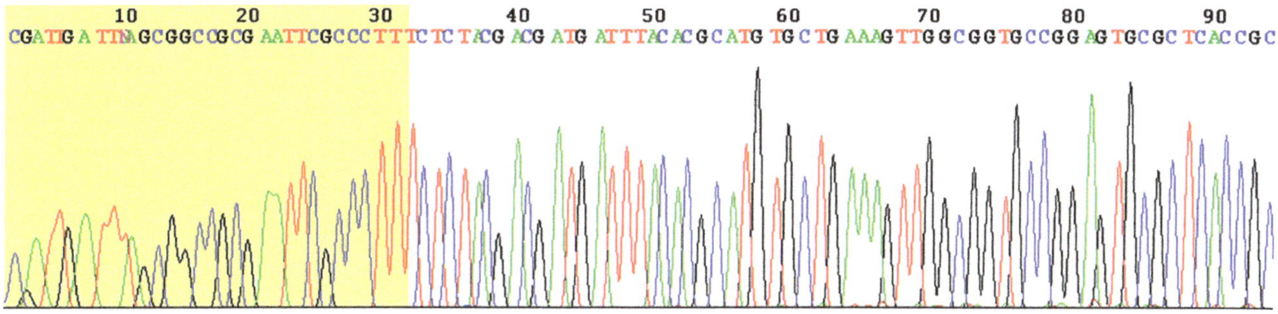

https://commons.wikimedia.org/wiki/File:Sanger_sequencing_read_display.png

NEXT GENERATION SEQUENCING ROLE-PLAY GAME

The purpose of the role-play is to demonstrate the technology that allows large numbers of unique DNA sequences to be sequenced simultaneously

Assign each member of the group one of the following jobs:

a. "Beads" – these hold unknown DNA strands to be sequenced - any number (example, for a group of 20 have 8 "beads")
b. "Reporters" – 1 for each "bead" (8 reporters for example group)
c. "Nucleotides" – 1-4 participants labeled "A", "C", "T", or "G" (1 of each base – if the group is small one participant can be more than one base)

Materials:

1. "Unknown" DNA sequences – as many as desired. 1 unique sequence per "bead" – each sequence has a "bar code" sequence at the beginning of the unknown sequence.
2. Nucleotide labels – "A" "T" "C" "G" – can have multiple sets
3. "Bar code" sequences that match "unknown"
4. Chart paper or whiteboard for Reporters
5. Markers

Procedure:

1. Give each "bead" an "unknown" DNA sequence with a barcode.

2. Have the beads distribute themselves around the classroom or meeting space.

3. Give each "reporter" a barcode. They should identify which "bead" sequence they match. They will be recording on the board or chart paper only the information from the matching barcode. Reporters stand to one side of the group – not next to the beads.

4. Give each Nucleotide a label – A, C, T, or G . These should be taped onto the nucleotide.

To Play:

A nucleotide (for example, "A") washes through and among the beads until each bead has been tagged. IF the bead determines that the nucleotide is the correct **base-pair match** for the FIRST position in their mystery sequence (the first base after the barcode), they call out "PING!" and their reporter marks an A on their board or chart. If there is no base-pair match, the "bead" stays silent. If there are two of the same matching base in a row, the bead calls out "PING PING" and the reporter records two bases.

5. Quickly, the next nucleotide (say, "G") washes through the group, tagging each bead. Again, if the G is the appropriate base-pair for the NEXT POSITION of their mystery sequence, the bead calls out "PING" and their reporter writes it on the board.

6. The remaining two nucleotides wash through the group and matching bases are called out by the beads.

7. The rotation of nucleotides begins again and continues as rapidly as the group can go until each of the "mystery sequences" has been recorded on the chart or whiteboard.

NEXT GENERATION SEQUENCING ROLE-PLAY GAME CONTINUED

8. The reporter should complete the double stranded sequence and then compare it to the mystery sequence for accuracy.

9. **SEE "NGSS Diagram"** for a pictorial description of the procedure.

Important reminders:
Nucleotides are only called in sequence – the beads may NOT "skip ahead" in their calling out. The "mystery sequence" is the **template** for the sequence being called out – if there is an "A" in the **mystery sequence** then the bead would call out "PING" when tagged by the "T" wash. "Beads" need to have a good understanding of base pairing.

1st rotation:	On board:
• During "A" wash, no bead calls out • During "G" wash, no bead calls out • During "C" wash, mystery bead AA calls out "PING" • During "T" wash, mystery bead AC calls out "PING	Barcode AA: C- Barcode AC: T-
2nd rotation: • During "A" wash, no bead calls out • During "G" wash, BOTH beads call out "PING" • During "C" wash, no bead calls out • During "T" wash, no bead calls out	Barcode AA: C-G- and Barcode AC: T-G-
3rd rotation: • During "A" wash, mystery bead AA calls out "PING" • During "G" wash, no bead calls out • During "C" wash, mystery bead AC calls out "PING-PING" • During "T" wash, no bead calls out	Barcode AA: C-G-A Barcode AC: T-G-C-C

Illustration:
"Mystery bead AC" – sequence AC-ACGGAC
"Mystery bead AA" – sequence AA-GCTCAG

Continue with rotations until all sequences have completed.

BARCODES FOR REPORTERS

AAAT	AATA	AGAA
AAAG	AAGA	ACAA
AAAC	AACA	AATT
AAAA	ATAA	AATG

SEQUENCES FOR BEADS

AAAT-AGTAGCATGACG

AAAC-GGATCGATCTAG

AAAG-TACTCGTACCAT

AAAA-CTCTGAGATCGA

AATA-GATTACGATACC

AACA-AATATGAGCGTC

AAGA-GAGATACTCTGC

ATAA-CATATACGTGTA

SEQUENCES FOR BEADS CONTINUED

AGAA-GGAGTTAGCCTA

ACAA-CCATTACACATT

AATT-GGGCCCATTTAT

AATG-AAATTAGGGCCC

Nucleotide Labels

A	T
C	G

"DATABASE" OF BACTERIAL DNA SEQUENCES

***E. coli*: AGTAGCATGACG**

***Salmonella*: GGATCGATCTAG**

***Bifidobacterium*: TACTCGTACCAT**

***Bacteriodes*: CTCTGAGATCGA**

***Lactobacillus*: GATTACGATACC**

***Peptococcus*: AATATGAGCGTC**

***Ruminococcus*: GAGATACTCTGC**

***Campylobacter*: CATATACGTGTA**

***Enterococcus*: AAATTAGGGCCC**

***Flavobacterium*: GGAGTTAGCCTA**

***Clostridium*: CCATTACACATT**

***Aeromonas*: GGGCCCATTTAT**

NGSS DIAGRAM (Explanation follows on next page)

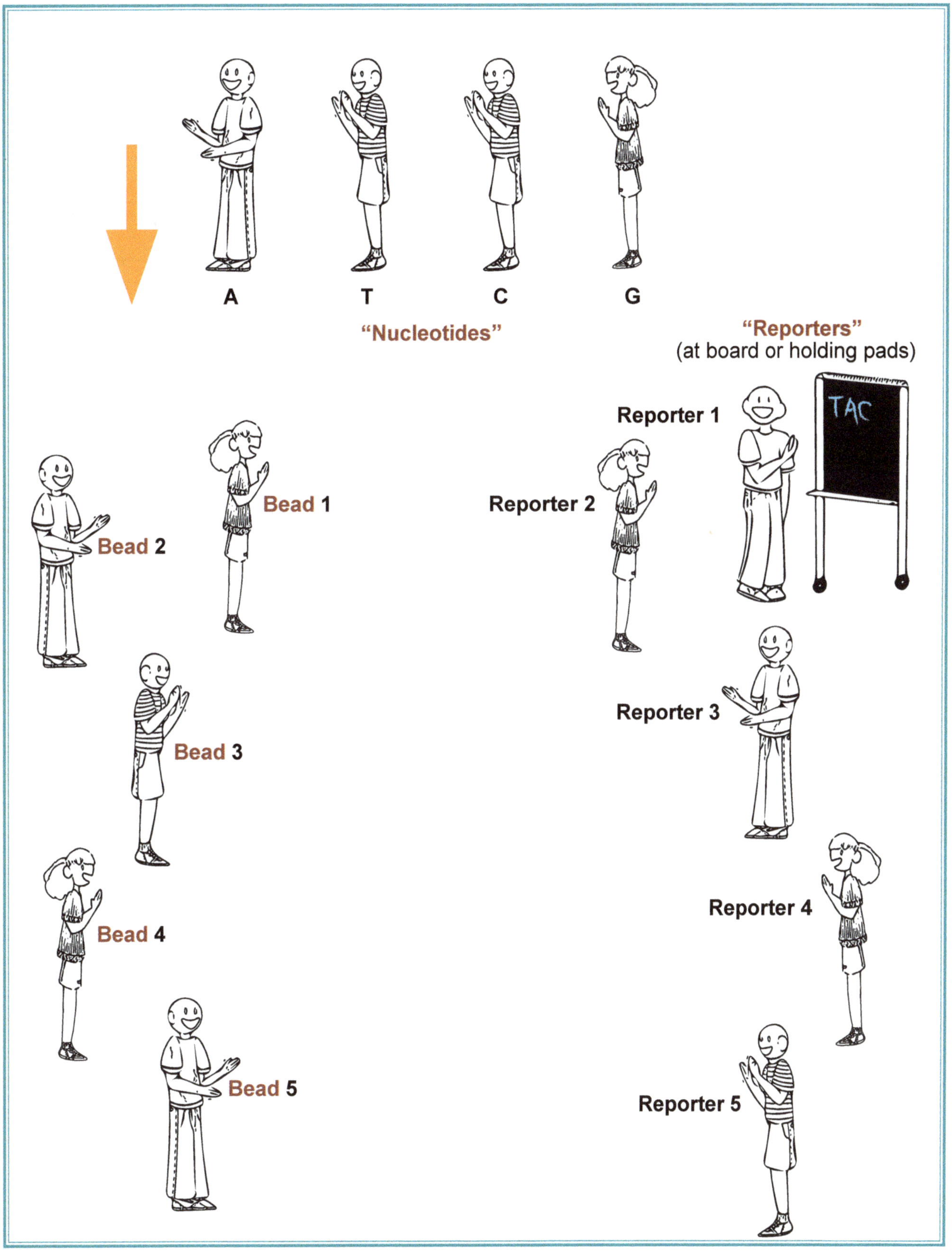

NGSS DIAGRAM CONTINUED

1. One at a time, each nucleotide base moves from one side of the room to the opposite, tagging each "bead".

2. **ONLY IF** the base PAIRS with the next base in the sequence that the Bead is holding, the BEAD says "PING" and that BASE's REPORTER writes the base on the whiteboard or chart. Each "PING" adds a base to the sequence the REPORTER is building.

3. Example:

Round 1:	• When base A tags him, he says nothing. (A does not base pair with A) • When base T tags him, he says "PING" and his reporter writes "T" on the board (T base pairs with A) • When base C tags him, he says nothing (C does not base pair with T) • When base G tags him he says nothing (G does not base pair with T)
Round 2:	• When base A tags him, he says "PING" and his reporter writes "A" on the board (A base pairs with T) • When base T tags him, he says nothing (T does not base pair with G) • When base C tags him, he says "PING" and his reporter writes "C" on the board (C base pairs with G) • When base G tags him, he says "PING PING" and his reporter writes "G G" on the board (G has base paired with the next two C's)

4. In each round, the four nucleotides pass in the same order (A-T-C-G), tagging each bead.

5. When all beads have been sequenced, the reporters should complete the double-stranded DNA sequence on their board/sheet.

For example, Reporter D should have on the board: T A C G G and then complete the sequence by writing A T G C C which will match the sequence that his/her partner Bead D was holding.

6. The activity can be made competitive by seeing which "BEAD-REPORTER" pair generates the most accurate sequence. Many times once the group gets the hang of the activity, they ask for another round to improve their speed and accuracy.

7. The activity can be used as a "Find the Culprit" extension by asking participants if any of the sequences represent a Salmonella presence. Use the "Database of Bacterial DNA Sequences" to compare the results.

Extracting DNA from Your Cells

Cells from the lining of your mouth come loose easily, so you will be able to collect cells containing your DNA by swishing a sports drink around in your mouth.

To extract DNA from your cells, you will need to separate the DNA from the other types of biological molecules in your cells. You follow the same basic steps that biologists use when they extract DNA (e.g. to clone DNA or to make a DNA fingerprint). You will follow these 3 easy steps to extract the DNA from your cheek cells.

1. Add Detergent **2. Add Enzymes** **3. Add Cold Alcohol**

Materials Needed:

- Bathroom cup (3-4 oz)
- Dish detergent
- Sports Drink (ex. Gatorade)
- Meat Tenderizer (Ex. Adolph's)
- 90% Isopropyl (rubbing) alcohol – COLD
  (from the freezer – take out just before use)
- Test tube
- Safety gloves
- Dropper
- Tall glass or test tube holder

Getting Your Sample of Cells

Fill about 1/3 of a bathroom cup with the sports drink. Swish the sports drink around in your mouth vigorously for at least one minute. You can gently scrape the insides of your cheeks with your teeth. You want to get plenty of cells. Then spit the drink back into the cup. Put on safety gloves before proceeding.

Procedure:

Step 1: Detergent

Add about 1 drop of detergent to a test tube. Now carefully pour the drink containing your cheek cells into the test tube with the detergent until the tube is half full.

> **What does the detergent do?**
> Cells are surrounded by cell membranes that are largely made up of lipids (or fat molecules). To open up the cell membranes, detergent is used to break up the lipids, just like they break up fats on dirty dishes.

Step 2: Enzymes

Add a pinch of the enzyme (meat tenderizer) to your test tube. With your gloved finger covering the top of the test tube, gently invert the tube five times to mix. Doing this gently is the key - you don't want to make the mixture foamy or break up the DNA strands. Let the mixture sit for at least 10 minutes. Prop it into a tall glass. You can discard your gloves into the trash.

> **What do the enzymes do?**
> The DNA in your cells is very long – almost 3 meters in each cell! To fit all that DNA into a cell, it is wrapped many times around proteins. The enzymes break down the proteins to allow the DNA to become free and unwind.

Extracting DNA from Your Cells continued

Step 3: Alcohol
Tilt the test tube slightly to one side. Using a dropper, slowly add COLD rubbing alcohol into the test tube; let the alcohol run down the side of the test tube so it forms a layer on top of the soapy liquid. Continue adding alcohol until you have about one inch (2 cm) of alcohol in the tube. Do not disturb the test tube for 10 minutes. DNA molecules will clump together where the salty, soapy mixture below meets the cold alcohol above, and you will be able to see tiny bubbles that, over time, will lift the wispy DNA strands up into the alcohol layer.

What does the alcohol do?
The cold alcohol reduces the ability of DNA to stay dissolved. When cold alcohol is gently poured on top of the cheek wash solution, the DNA separates into the alcohol layer, while the lipids and proteins stay in the solution.

Adapted from the Genetic Science Learning Center's "How to Extract DNA from Any Living Thing"
http://learn.genetics.utah.edu/content/labs/extraction/howto/

DNA SEQUENCING BY CHAIN TERMINATION –
SEQUENCE THE MYSTERY DNA!

Newly Synthesized Strands: cut out each one

You will have 16 strands of different lengths, ending with a different base. Tape these to the appropriate box on your handout.

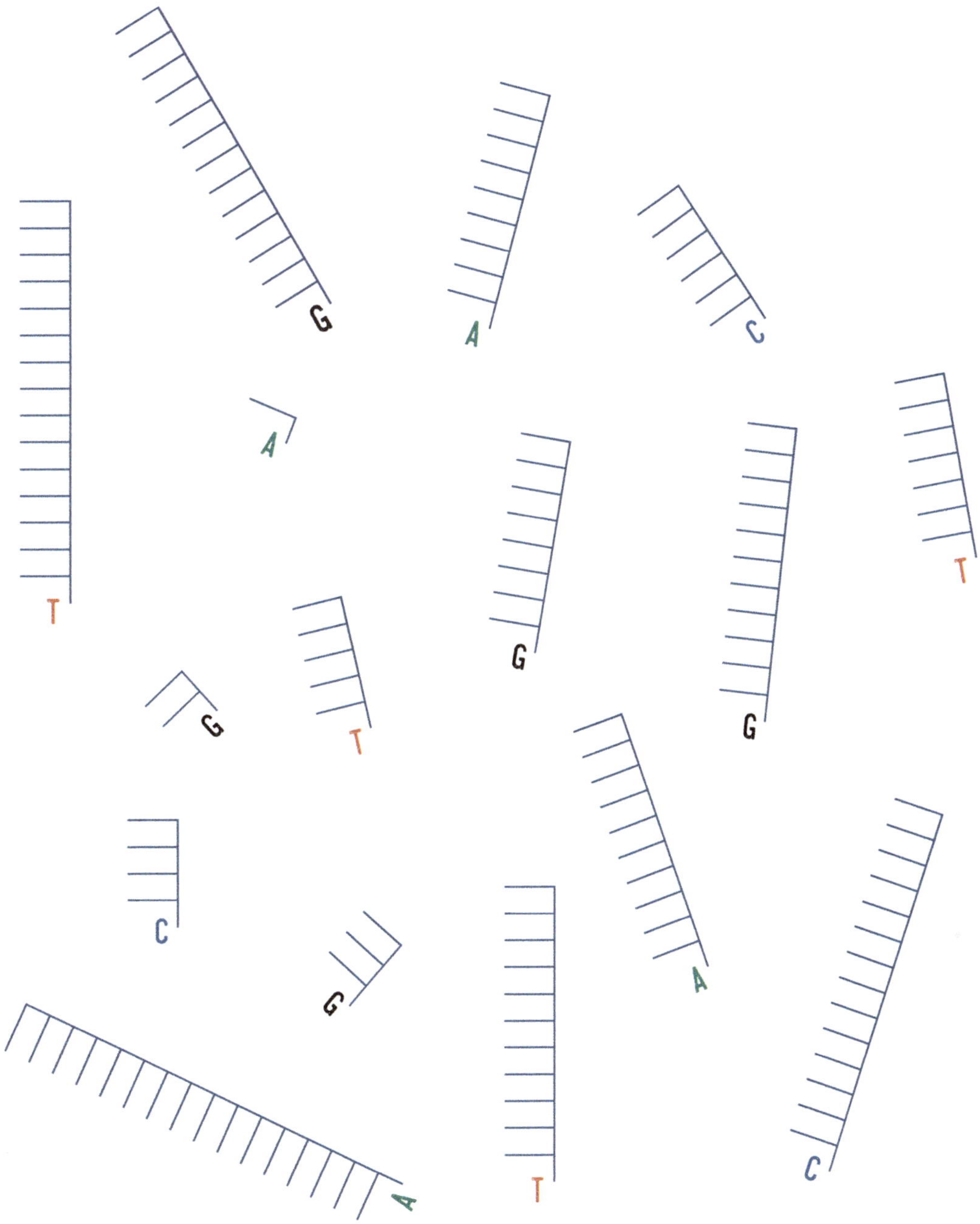

SANGER SEQUENCING SIMULATION INSTRUCTIONS

1. Cut out all 16 “Mystery DNA” sequences. (Cut closely so they will fit in the boxes below.)
2. In each box, match up a “Mystery sequence” with a newly synthesized sequence ending with the requested base (“A”, “C”, “G”, or “T”)
3. The pattern in the boxes should look like:

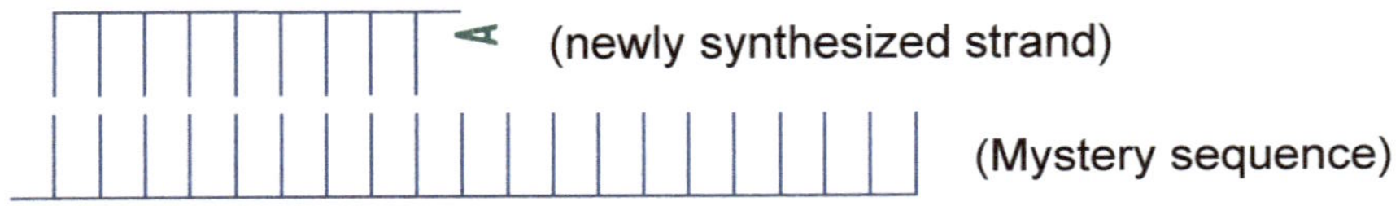

All Sequences Ending with “A”

All Sequences Ending with “C”

All Sequences Ending with “G”

All Sequences Ending with “T”

4. Starting from the SHORTEST fragment to the LONGEST, read the sequence of the final bases and record it here:

____ ____ ____ ____ ____ ____ ____ ____ ____ ____ ____ ____ ____ ____ ____ ____

LESSON 5 1, 2, 4, GROW!

HOW BACTERIAL GROWTH IS AFFECTED BY ENVIRONMENTAL CONDITIONS

Introduction

When bacteria grow, they do so by splitting in half. Each cell becomes two cells. This is called **binary fission**. Unlike human cells or other eukaryotes, bacterial cells act independently as they multiply. This means that you can get an extremely large number of bacteria fast! This activity is designed to model how bacteria grow and to show how simple everyday conditions can affect how and where bacteria grow in and on the human body. In addition, it will show how to control the growth of bacterial colonies. The five bacteria featured in the 1, 2, 4 Grow! game are *Lactobacillus casei, Helicobacter pylori, Salmonella enteritidis, Escherichia coli,* and *Staphylococcus aureus* (see the presentation for details). Most of these are native to the human body. Under certain conditions, several of these bacteria can cause illness or injury, especially if they are left to duplicate uncontrolled.

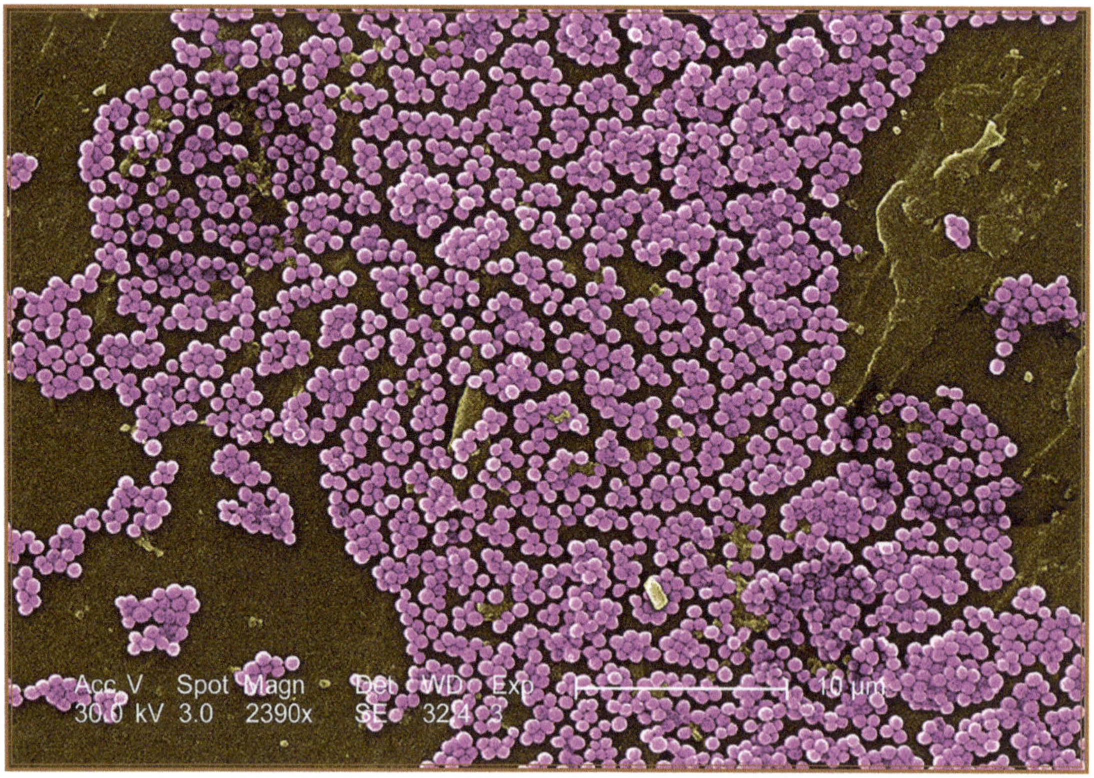

Scanning electron microscopic view of MRSA. https://phil.cdc.gov/phil/details.asp

Lesson 5 1, 2, 4, Grow!

Key Teaching Points

1. Bacteria multiply by binary fission.
2. Populations of bacteria can grow quickly, but they also die easily.

Learner Outcomes

- Understand how bacterial populations grow.
- Construct and interpret population data on a graph.
- Predict how a population will react to a given stress.

Skill Level: Advanced, ages 14-18

Success Indicators

- Participants will plan a set of starting conditions to match a set of ending conditions.

Life Skills

Keeping Records, Critical Thinking

Time Needed

90-120 minutes

Space

Classroom setting with a table for each participant group

Suggested Group Size

2-3 participants per group

Materials List

1. One complete game set for each participant group. Game sets are comprised of:
    - One game board
    - One Scenario Sheet
    - One set of the Event Cards (print one sheet per group)
2. Activity Data Sheets (1 per group per scenario)
3. Five differently colored pens or pencils or markers to record the bacterial growth on the Activity Data Sheets

Options for the game board:

Participants will play this game through four or five different scenarios. Choose one of the following options for game play and prepare your materials accordingly:

- **Dry erase**: Participants will make dots or tally marks on a dry erase board for the bacteria that are added or subtracted in each scenario.
    - Materials needed: one dry erase board per group, and an assortment of dry erase markers
- **Laminated game board**: Participants will use dry erase markers to make dots or tally marks directly on the laminated game board for the bacteria that are added or subtracted in each scenario.
    - Materials needed: one laminated game board per group and an assortment of dry erase markers
- **Multiple game boards**: Participants will use a different game board for each scenario of the game and make marks directly on the game board for the bacteria that are added or subtracted in each scenario.
    - Materials needed: one printed game board PER SCENARIO, per group (4-5 game boards per group)

LESSON 5 1, 2, 4, GROW!

Before the Activity

1. Review the presentations for this lesson to become familiar with the content.
2. Print one page of Event Cards for each group. [Note: Either cut these out before the activity or provide scissors to each group for participants to cut them out.]
3. Prepare 5 data sheets, one 1, 2, 4 Grow! Scenario List, and one or more game boards for each group.
4. Play each scenario in the game to familiarize yourself with the learning experience.
5. Discuss and review any questions about bacteria before starting. The facilitator should remind the participants they are in the role of "scientist" and should be collecting data. Use the questions to direct participants to explore the cards and the different bacteria on them.

The game has five rules:

1. Record the number of bacteria present on the data sheet at the beginning of each scenario.
2. Play an event card.
3. Record the effect of the Event on the bacteria on the Data Sheet and calculate the new number of bacteria.
4. Double all remaining bacteria at the end of the round.
5. Continue until all five rounds of each scenario have been completed.

The game should play in 5-10 minutes for each scenario. Each scenario will consist of 5 rounds (generations) where players will draw an event, do what it says on the card, and double all bacteria at the end of each round.

Opening Question:

How do you think bacteria grow?

Follow up question:

If bacteria split in two after every generation, how many bacteria are present after one, two, or four generations? [Note: this question should be a review for participants from previous lessons.]

Let's Do It!

1. Begin by asking participants the Opening Question.
    - How do you think bacteria grow?
2. Elicit participant responses, both correct and incorrect. Record responses where all participants can see them and build from them to explain to the participants that bacteria grow by splitting in half in

a process called binary fission. Remind the participants that all cells, including their own, divide in ways similar to that of bacteria. However, plant and animal cells are not independent microorganisms, but form tissues in a way that bacteria do not.

3. Ask the Follow-up question.

    - If bacteria split in two after every generation, how many bacteria are present after one, two, or four generations? [Note: this question should be a review for participants from previous lessons.]

4. Participants should be able to respond with gen 1 = 2, gen 2 = 4, and gen 4 = 16. Hook the participants into the game by having them play one round using five "All Quiet on the Western Front" cards and beginning with one Salmonella. Optionally, participants can act as bacteria and "double" after each round.

5. With each station set up at tables, provide each group with the scenarios list. Ask participants to work through scenarios 1-5 and record and graph the results on the data sheets. Each bacteria should have a different color line on the graph to be easily read. You may have participants play through a scenario and discuss or have participants play through all scenarios and compare their results.

    1. Play scenario 1 "Stomach Conditions"
    2. Play scenario 2 "Skin Conditions"
    3. Play scenario 3 "Mouth Conditions"
    4. Play 4th and 5th rounds with random events (participants can shuffle the event card deck)

6. Ask the Share questions.

    - Compare the graphs of each group. Scenarios 1, 2, and 3 should have identical data. What differences do you notice amongst the graphs?
    - Guide participants to discuss the differences in scenarios 4 and 5. What makes graphs 4 and 5 different for each group?

7. Ask the Reflect questions.

    - In scenario 1, what could have been a source for the explosive bacterial growth during turn two? Did you wipe out a species completely?
    - In scenario 2, what kept the populations from growing in relation to the environment?

- In scenario 3, why do the bacterial populations bounce back in the mouth?
- How can the growth of bacteria be controlled?

8. Ask the Generalize questions.

- What conclusions can you draw about bacterial growth?
- What are some ways we can use bacterial growth to help us?

9. Ask the Apply question.

- Research ways in which bacteria growth can be beneficial. See the Learn More section for options.

Talk It Over

SHARE 1	REFLECT 2
Compare the graphs of each group. Scenarios 1,2, and 3 should have identical data. What differences do you notice amongst the graphs? Guide participants to discuss the differences in scenarios 4 and 5. What makes graphs 4 and 5 different for each group?	In scenario 1, what could have been a source for the explosive bacterial growth during turn two? Did you wipe out a species completely? In scenario 2, what kept the populations from growing in relation to the environment? In scenario 3, why do the bacterial populations bounce back in the mouth? How can the growth of bacteria be controlled?
GENERALIZE 3	**APPLY 4**
What conclusions can you draw about bacterial growth? What are some ways we can use bacterial growth to help us?	Research ways in which bacteria growth can be beneficial. See the Learn More section for options.

Term and Concept Discovery

Aerobes: organisms that cannot live without oxygen
Anaerobes: organisms that can live without oxygen
Binary fission: the process of a bacterium splitting and becoming two
Facultative anaerobe: organisms that can live with or without oxygen
Growth curve: a graph of population size over time

Variations

For assessing the concept of bacterial growth, participants can challenge other groups to meet certain conditions after a given amount of turns and then have the challenged group plan out how to reach their goals by constructing an event deck. Make sure that ending numbers are attainable by choosing ending numbers of 1, 2, 4, 8, 16, or 32.

For Example:

"Make a population of lactobacillus that starts with 4 bacteria and ends with 32 bacteria after turn 5"

The answer to which would be to arrange the events as follows:

1. Cefazolin or Exposed to the Elements (reduce the lactobacillus count to 1 then double the remaining one for 2)
2. All Quiet on the Western Front (2 goes to 4)
3. All Quiet on the Western Front (4 goes to 8)
4. All Quiet on the Western Front (8 goes to 16)
5. All Quiet on the Western Front (16 goes to 32)

LEARN MORE

http://www.cellsalive.com

https://asm.org/Podcasts/BacterioFiles

How bacteria can help plants:
https://youtu.be/ylSrMwpH3GI

Good bacteria in your gut:
https://youtu.be/eg6Uy18qpR8

FDA food poisoning list by species and lag time before symptoms:
http://www.fda.gov/food/resourcesforyou/consumers/ucm103263.htm

Education Standard(s)

HS-LS2-2. Use mathematical representations to support and revise explanations based on evidence about factors affecting biodiversity and populations in ecosystems of different scales.
HS-ETS1-3. Evaluate a solution to a complex real-world problem based on prioritized criteria and trade-offs that account for a range of constraints, including cost, safety, reliability, and aesthetics, as well as possible social, cultural, and environmental impacts.
CCSS. Math.Content.HSS-IC.A.1 Understand statistics as a process for making inferences about population parameters based on a random sample from that population.

Tags

Populations, graphing, exponential growth

Lesson 5 Print Materials

Facilitator Print Materials

None

Group Handouts

Game boards - one per group or one per group per scenario (see Before the Activity for options)
Event Cards
1, 2, 4 Grow! Scenario List
Data Sheet

Youth Handouts

None

See page 3 for access to digital assets & extra enrichment.

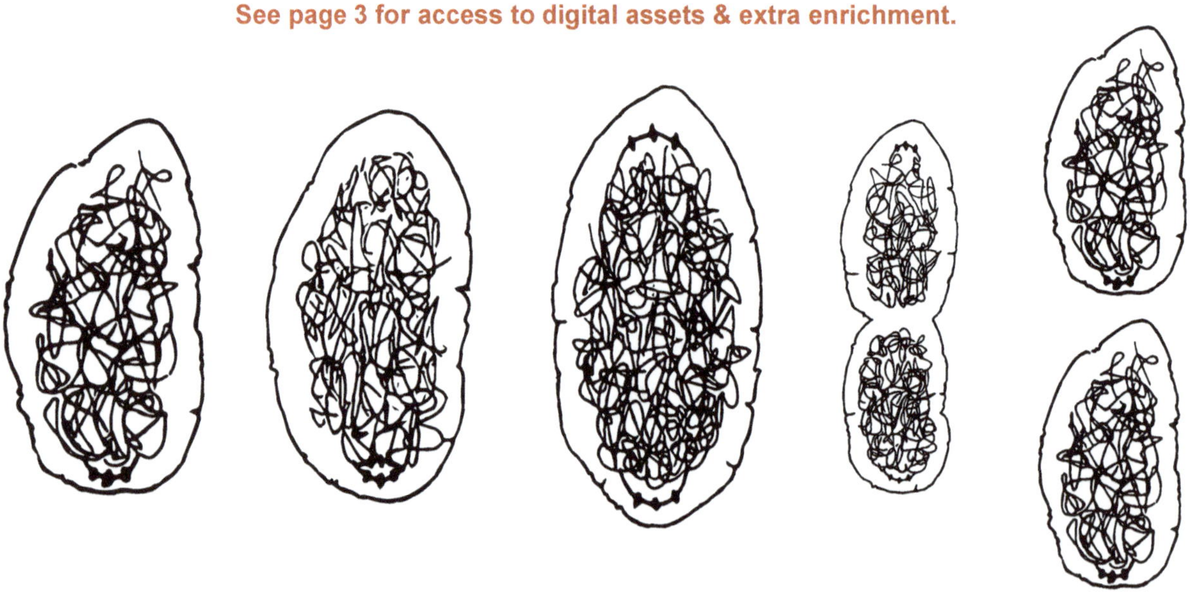

Populations of bacteria multiply through a process called binary fission. Individual cells are split in half to form independent, new cells. Because of this growth, bacterial populations can multiply quickly, but they are also prone to environmental stress. In this lesson, you will learn about bacterial growth and interpret growth trends under different scenarios.

1, 2, 4, GROW! GAME

Rules:
1. Record the number of bacteria present on the data sheet at the beginning of each scenario.
2. Play an event card.
3. Record the effect of the Event on the bacteria on the Data Sheet and calculate the new number of bacteria.
4. Double all remaining bacteria at the end of the round.
5. Continue until all five rounds of each scenario have been completed.

PICK UP

DISCARD

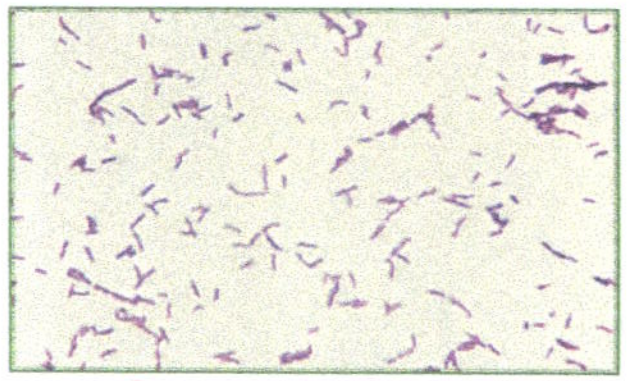

LACTOBACILLUS
Gram positive
Aerobic/anaerobic
Rod-shaped
Non-motile
No spores

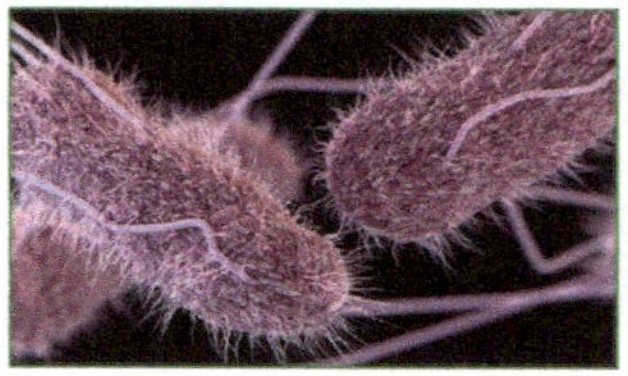

SALMONELLA
Gram negative
Aerobic/anaerobic
Rod-shaped
Motile
No spores
Cannot ferment lactose

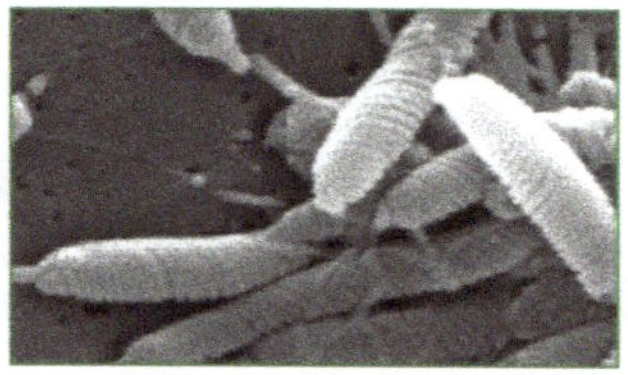

HELICOBACTER
Gram negative
Aerobic
Spiral-shaped
Motile
No spores

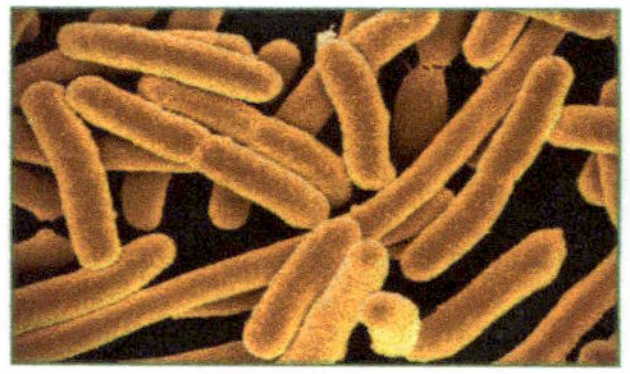

E. COLI
Gram negative
Aerobic/anaerobic
Rod-shaped
Motile
No spores
Can ferment lactose

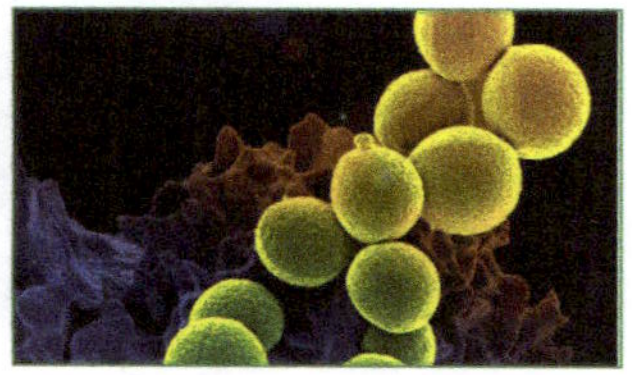

STAPHYLOCOCCUS
Gram positive
Aerobic/anaerobic
Spherical
Non-motile
No spores
Occur in clumps

EVENT CARDS

Sugar and spice, not everything is nice - double all bacteria (make sure to double again at the end of the round)

Exposed to the Elements:
3 cards per page
Kill ¼ of all bacteria
(round down)

Exposed to the Elements:
3 cards per page
Kill ¼ of all bacteria
(round down)

Exposed to the Elements:
3 cards per page
Kill ¼ of all bacteria
(round down)

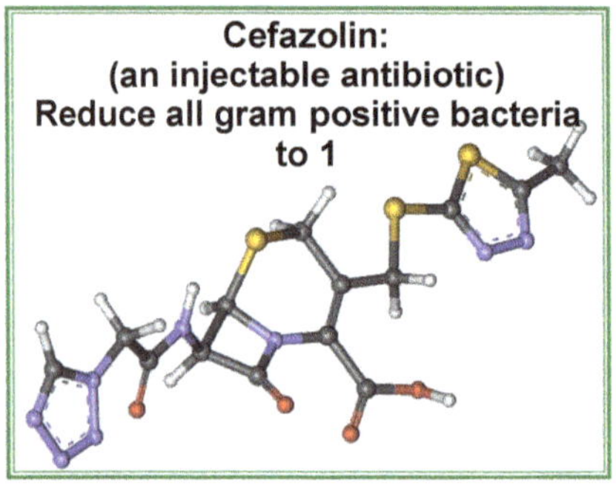
Cefazolin:
(an injectable antibiotic)
Reduce all gram positive bacteria to 1

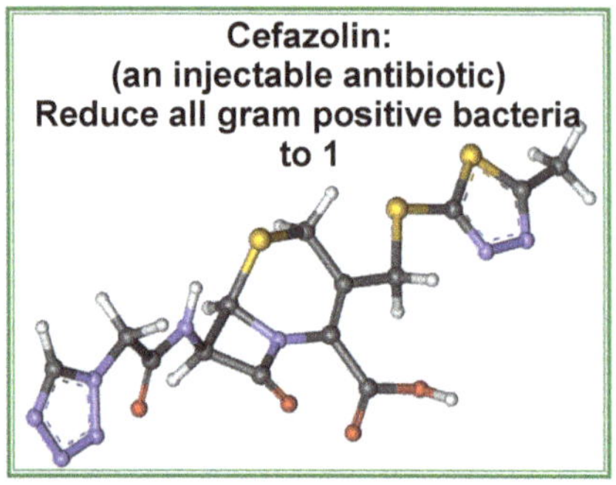
Cefazolin:
(an injectable antibiotic)
Reduce all gram positive bacteria to 1

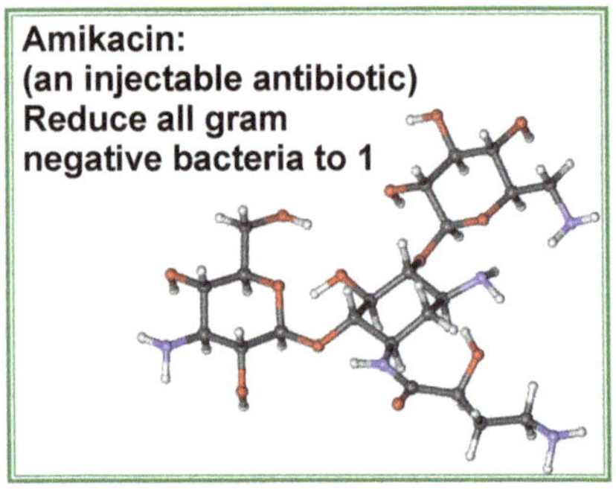
Amikacin:
(an injectable antibiotic)
Reduce all gram negative bacteria to 1

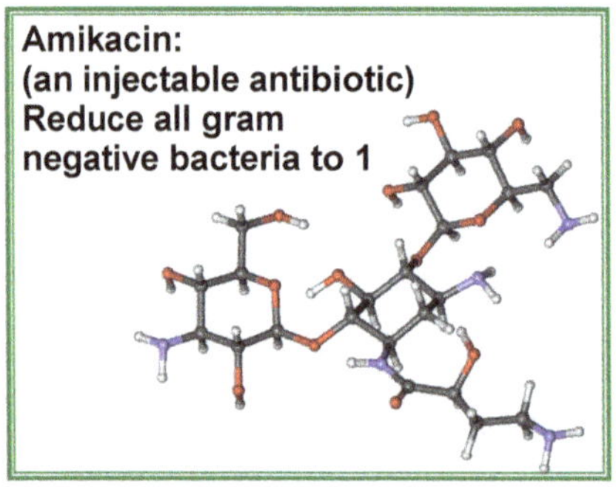
Amikacin:
(an injectable antibiotic)
Reduce all gram negative bacteria to 1

All quiet on the Western Front:
Normal conditions apply

All quiet on the Western Front:
Normal conditions apply

All quiet on the Western Front:
Normal conditions apply

All quiet on the Western Front:
Normal conditions apply

All quiet on the Western Front:
Normal conditions apply

All quiet on the Western Front:
Normal conditions apply

All quiet on the Western Front:
Normal conditions apply

EVENT CARDS EVENT CARDS EVENT CARDS
EVENT CARDS EVENT CARDS
EVENT CARDS EVENT CARDS EVENT CARDS
EVENT CARDS EVENT CARDS
EVENT CARDS EVENT CARDS EVENT CARDS
EVENT CARDS EVENT CARDS
EVENT CARDS EVENT CARDS EVENT CARDS
EVENT CARDS EVENT CARDS
EVENT CARDS EVENT CARDS EVENT CARDS
EVENT CARDS EVENT CARDS
EVENT CARDS EVENT CARDS EVENT CARDS
EVENT CARDS EVENT CARDS
EVENT CARDS EVENT CARDS EVENT CARDS
EVENT CARDS EVENT CARDS
EVENT CARDS EVENT CARDS EVENT CARDS
EVENT CARDS EVENT CARDS
EVENT CARDS EVENT CARDS EVENT CARDS

1, 2, 4 GROW! SCENARIO LIST

Before each scenario, stack the cards in the appropriate order and place them on the Pick-Up pile on the game board.

Scenario 0:
This is your trial run! Follow along with the facilitator as this round is played.

Starting Bacteria:	Salmonella x 1
Event Card Order:	All Quiet on the Western Front (normal conditions) x 5

Scenario 1:
In this scenario, the event cards will be organized to simulate your gut as you start with helicobacter pylori, multiply to maximum (which can lead to gut ulcers), and then administer a gram-negative antibiotic to kill it.

Starting Bacteria:	Salmonella x 1 Lactobacillus x 1 Helicobacter x 1 Staphylococcus x 1 E. coli x 1
Event Card Order:	All Quiet on the Western Front (normal conditions) Sugar and Spice Not Everything is Always Nice (doubles all bacteria) All Quiet on the Western Front (normal conditions) All Quiet on the Western Front (normal conditions) Amikacin applied (reduce all gram-negative bacteria to 1)

Scenario 2:
This scenario takes place on the skin. The Event Cards will simulate the growth of E. coli and Staphylococcus bacteria on the skin (which can lead to boils) until Amikacin is applied. However, that antibiotic does not stop the growth of a gram-positive bacteria; thus, infection is still possible if the wrong antibiotic is applied.

Remember! You cannot kill a fraction of a bacterium. If you find yourself with a partial bacterium, kill the entire microbe.

Starting Bacteria:	E. coli x 4 Staphylococcus x 4
Event Card Order:	Exposed to the Elements (kill ¼ of all bacteria) Exposed to the Elements (kill ¼ of all bacteria) Exposed to the Elements (kill ¼ of all bacteria) Amikacin Applied (reduce all gram negative bacteria to 1) All Quiet on the Western Front (normal conditions)

1, 2, 4 GROW! SCENARIO LIST CONTINUED

Scenario 3:
This scenario takes place in the mouth. When Lactobacillus is able to grow unchecked in the mouth, it can cause tooth decay and cavities. Notice also that if antibiotics are applied with low numbers of bacteria present, numbers of bacteria can multiply rapidly.

Starting Bacteria:	Lactobacillus x 1 Streptococcus x 1
Event Card Order:	Cefazolin (reduce all gram positive bacteria to 1) All Quiet on the Western Front (normal conditions) All Quiet on the Western Front (normal conditions) Sugar and Spice Not Everything is Always Nice (doubles all bacteria) All Quiet on the Western Front (normal conditions)

Scenarios 4 and 5:
Shuffle all of the Event Cards. Without looking, choose five to create your own random, unique scenario. Continue to use the game board and a Data Sheet to record and graph your data.

1, 2, 4, GROW! DATA SHEET

Scenario Name: ______________________

Key	
E. Coli	
Staph	
Lacto	
Helicobacter	
Salmonella	

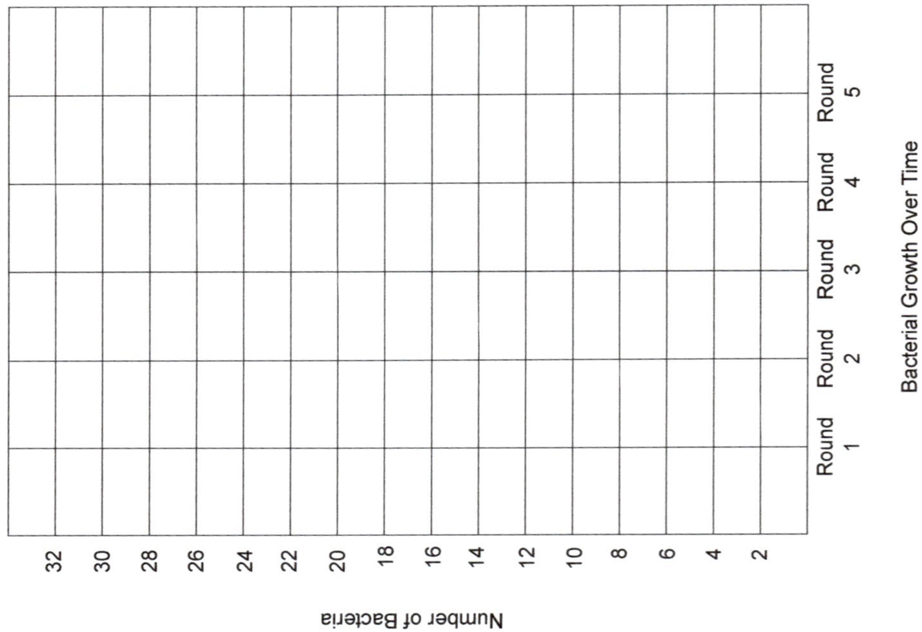

	Beginning number	effect of event	new number	double-end of round	Beginning number- Round 2	effect of event	new number	double-end of round	Beginning number- Round 3	effect of event	new number	double-end of round	Beginning number- Round 4	effect of event	new number	double-end of round	Beginning number- Round 5	effect of event	Ending number
E. Coli				× 2				× 2				× 2				× 2			
Staph				× 2				× 2				× 2				× 2			
Lacto				× 2				× 2				× 2				× 2			
Helico				× 2				× 2				× 2				× 2			
Salmonella				× 2				× 2				× 2				× 2			

Lesson 6 ✹ Double, Double, Toil, and Trouble

Using Mathematical Tools in the Microbiology Lab

Introduction

Important algebraic skills are needed for measuring objects within the field of microbiology. In this lesson, participants will be expected to perform two different tasks that will teach them skills used in the microbiology lab today. These tasks are framed around a resurgence of the black plague and allow participants to experience some of the same procedures used by microbiologists to isolate and deliver cures for disease-causing bacteria. The skills presented here can be applied to labs and science experiences outside of the lesson, making this a process with an applicable purpose. Without these techniques, the discoveries in the other lessons would not be possible.

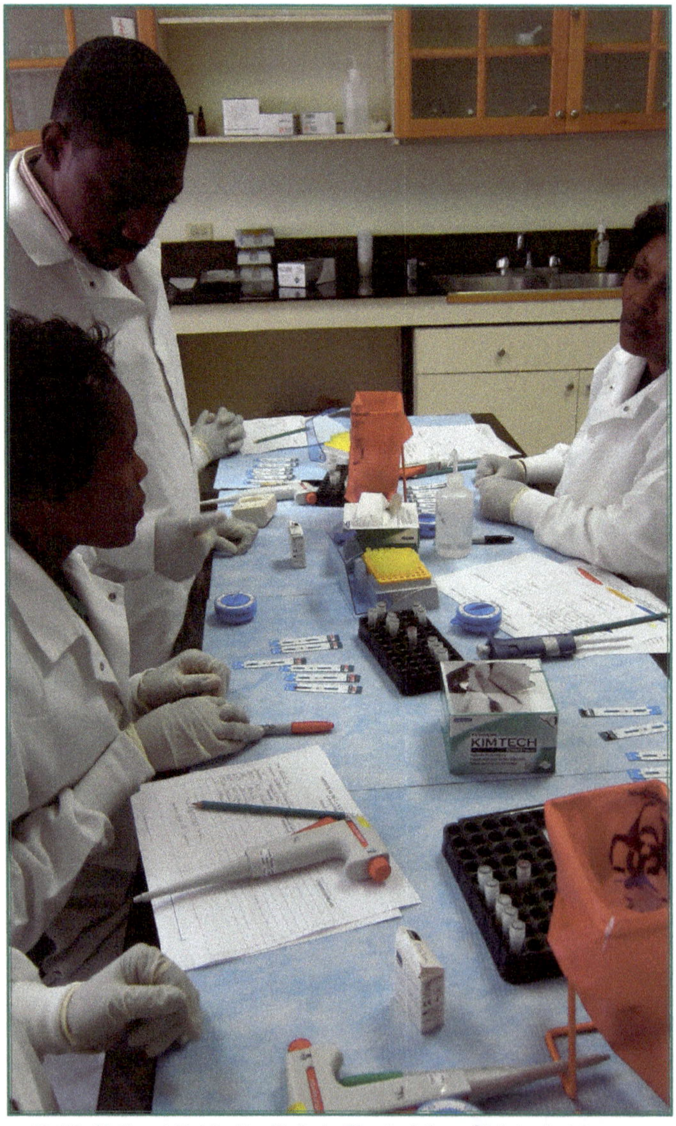

Haiti's National Public Health Lab. The training of lab technicians on laboratory techniques and methods for accurately recording data. phil.cdc.gov

Lesson 6 ✱ Double, Double, Toil, and Trouble

Key Teaching Points

1. Lab technicians use math and measurement to perform tasks in the lab.
2. A solution is a mixture in which a solute is dissolved in a solvent.
3. Serial dilution can be used to adjust the concentration of a solution.
4. A cell viability count can be used to determine how well a cell population will function.

Learner Outcomes

- Understand the process and uses of performing a serial dilution.
- Calculate cell viability of a given sample.

Skill Level: Advanced, ages 14-18

Success Indicators

- Participants will perform a serial dilution accurately.
- Participants will count a sample of cells in 2 minutes or less.

Life Skills

Job Readiness, Critical thinking, Sharing

Time Needed

90-120 minutes

Space

2'x4' counter top

Suggested Group Size

2-3 participants per group

Materials List

- Data Sheets (1 per group)
- Red Drink Mix (premixed drink mix)
- 100 mL Graduated cylinder (one per group)
- 2 clear plastic cups per participant
- 10 clear plastic cups per group
- 4 "fun size" M & Ms packages per group
- Pitcher (mixing solutions)
- Digital scale
- Spoon (1 per participant)
- Paper towels

Please be aware of any food allergies participants may have and take precautions as necessary.

Lesson 6 ✱ Double, Double, Toil, and Trouble

Before the Activity

- Review the presentation for this lesson to become familiar with the content.
- Prepare technology to show the presentation and a video to the participants.
- Familiarize yourself with using a graduated cylinder.
- Prepare copies of the Data Sheets for each group.
- Prepare the materials for Serial Dilution and Cell Viability Counting for each group.
- Prepare two clear plastic cups per participant, one with red powdered drink mix, and one with water.

Part One

Opening Question:

Describe the differences and the similarities between the powder and the liquid.

Follow up question:
(for after the drink mix and water have been combined)

What are the differences now?

Let's Do It!

1. Give participants one spoon, one cup containing powdered red drink mix, and a cup of water. Ask them to sample a little of both.
2. Ask the Opening Question.
    - Describe the differences and the similarities between the powder and the liquid.
3. Participants next add the powder to the water, stir, and sample the red drink mixture. Ask the Follow-up question.
    - What are the differences now?
4. Present the lesson presentation to the participants. Point out that the drink mix and water created a solution. The presentation may be delivered all at once or may be divided into Part One and Part Two.
5. Announce that a deadly bacteria has come back - the black plague!
6. Show the following video to introduce or review plague: https://youtu.be/SQBxTLjYRsl

LESSON 6 ✹ DOUBLE, DOUBLE, TOIL, AND TROUBLE

7. Explain to the participants that they must isolate the bacterium for study. The only way to isolate the bacteria is to make a nano (0.000000009) scale solution, but they only have a graduated cylinder. TIME FOR A SERIAL DILUTION!!!!!

8. Have participants measure out 9 g of powdered red drink mix for the stock solution of 100 mL of water in cup 1. Fill the cups labeled 2-10 with 90 mL of water each to start. Then, by following the directions below, finish mixing the solutions for cups 2-10. Notice the solution with the measurement 0.000000009. This is your nano-scale solution of red drink mix.

Note: You must mix the previous solution first before taking 10 mL and moving it to the next cup.

Part 1 Procedure: Serial Dilutions

1. Label each cup with one number, 1-10.
2. Use the graduated cylinder to measure 100 mL of water into Cup 1.
3. Use the graduated cylinder to measure 90 mL of water into Cups 2-10.
4. Use the digital scale to measure 9 g of red drink mix. Add it to Cup 1 and stir until dissolved.
5. Use the graduated cylinder to measure 10 mL from Cup 1 and add it to Cup 2. Stir for 10 seconds.
6. Use the graduated cylinder to measure 10 mL from Cup 2 and add it to Cup 3. Stir for 10 seconds.
7. Repeat until you have stirred 10 mL of solution into every cup from the previous cup.
8. Calculate the concentration of all 9 of the cups based on the fact that 90 mL was water and 10 mL was red drink mix from the previous cup. (M1V1=M2V2)
9. Participants should recognize the pattern in the results for the equations after the first couple of calculations.

Encourage participants to extrapolate the molarity of the remaining solutions once they have discovered the pattern. Check for proper decimal placement.

Example math:

$$M_1 \cdot V_1 = M_2 \cdot V_2$$

$$9M \cdot 10mL = ?M \cdot 100mL$$

$$\frac{90M \cdot \cancel{mL}}{100\cancel{mL}} = \frac{?M \cdot \cancel{100mL}}{\cancel{100mL}}$$

$$.9M = ?M$$

M= molarity
V= volume

10. Ask the Share questions.

    - How many dilutions did it take to get down to the nano (0.000000009) level with the red drink mix (bacteria)?
    - Can you taste any red drink mix flavor after diluting this much? How do you know the red drink mix is present?

11. Ask the Reflect questions.

    - What pattern do you notice in the different concentrations of red drink mix?
    - How much powder would it take to make a 3 M solution in 100 mL?
    - What is another way to produce very dilute solutions? Is it practical?

12. Ask the Generalize questions.

    - What would happen to the taste of red drink mix if the volume were doubled but the amount of powder remained the same?
    - If you were diluting by a factor of 2 instead of 10, what would the concentration be after the third dilution (start with a 1M solution)?

13. Ask the Apply question.

    - Think of household solutions such as bleach, isopropyl alcohol, hydrogen peroxide, etc. See how many you can find in your house and make a poster ranking them from high to low concentration by percent.

This is a good time to break if needed.

Part Two

1. If a new class has begun, review serial dilution and the purpose for isolating a small enough sample of the plague bacteria to be able to isolate and study the microorganism.

2. Now you have the sample of isolated bacteria and need to find out how well the pathogen stays alive after being in the air for 24 hours for study. You must do a cell viability count!

3. A **cell viability** count is a method of calculating the ratio of live cells to all cells in a sample. With this method, scientists can project how long a cell sample will last in a lab.

Part Two Procedure: Cell Viability Counting

1. Begin this activity with four bags of M&M's per group.
2. Each group will open one fun-size packet of M & M's and put the other bags to the side for now.
3. Set a stopwatch or timer to 1 minute.
4. As the timer is started, participants will spread out the M&M's. Choose and announce a color of M&M to identify and count. This color represents viable (living) cells.
5. Participants will count the total amount of M&M's and also the number of the color that represents viable cells.
6. Participants will record the total number of M&M's and the number of M&M's that are viable on the data sheet.

7. Steps 1-6 will be repeated three times, once with each of the other bags of M&M's.
8. Participants will add the total number of M&M's for all four packs of M&M's.
9. Participants will add the total number of viable M&M's for all four packets of M&M's.
10. To calculate the percentage of live cells, the total number of viable cells will be divided by the total number of cells. Multiply the quotient by 100 to convert the decimal to a percent.

$$\frac{\text{15 viable “cells”}}{\text{57 total “cells”}} = .26$$

$$.26 \times 100 = 26\%$$

11. Ask the Share questions.

    - What does percent viability tell you? Does your group's percent viability match other group's viability? Why or why not?

12. Ask the Reflect questions.

    - What do you think a high rate of viability is?
    - Does low viability mean the population of bacteria will disappear? Think back to the 1, 2, 4 Grow! game.

13. Ask the Generalize questions.

- What applications of viability counts could be helpful with organisms other than bacteria?

Talk It Over

SHARE 1	REFLECT 2
Part 1: How many dilutions did it take to get down to the nano (0.000000009) level with the red drink mix (bacteria)? Can you taste any red drink mix flavor after diluting this much? How do you know the red drink mix is present? **Part 2:** What does percent viability tell you? Does your group's percent viability match other groups viability? Why or why not?	**Part 1:** What pattern do you notice in the different concentrations of red drink mix? How much powder would it take to make a 3 M solution in 100 mL? What is another way to produce very dilute solutions? Is it practical? **Part 2:** What do you think a high rate of viability is? Does a low viability mean the population of bacteria will disappear? Think back to the 1, 2, 4 Grow! game.
GENERALIZE 3	**APPLY 4**
Part 1: What would happen to the taste of red drink mix if the volume were doubled but the amount of powder remained the same? If you were diluting by a factor of 2 instead of 10, what would the concentration be after the third dilution (start with a 1M solution)? **Part 2:** What applications of viability counts could be helpful with organisms other than bacteria?	Think of household solutions such as bleach, isopropyl alcohol, hydrogen peroxide, etc. See how many you can find in your house and make a poster ranking them from high to low concentration by percent.

LESSON 6 DOUBLE, DOUBLE, TOIL, AND TROUBLE

Term and Concept Discovery

Cell Viability: what percentage of a sample of organisms are alive

Concentration: a measurement for the strength of a solution

Dilute: the act of making a solution less concentrated

Molarity: the SI unit for concentration

Serial Dilution: a process by which laboratories can make very accurate small-scale solutions

Solute: the substance dissolved in a solution

Solution: a homogeneous, molecular mixture of two or more substances

Solvent: the substance into which a solute is dissolved in a solution

Variations

Have participants mix their own solutions for part one. Pick a concentration and let participants mix it.

LEARN MORE

Step-by-step for a serial dilution:
http://www.wikihow.com/Do-Serial-Dilutions

Education Standard(s):

HS-LS2-1. Use mathematical and/or computational representations to support explanations of factors that affect carrying capacity of ecosystems at different scales.

CCSS.Math.Content.HSA-CED.A.2 Create equations in two or more variables to represent relationships between quantities; graph equations on coordinate axes with labels and scales.

CCSS.Math.Content.HSA-CED.A.4 Rearrange formulas to highlight a quantity of interest, using the same reasoning as in solving equations. For example, rearrange Ohm's law V = IR to highlight resistance R.

Tags

Measuring, diluting, counting, percentages

Lesson 6 Print Materials

Facilitator Print Materials
None

Group Handouts
None

Youth Handouts
Data sheet

See page 3 for access to digital assets & extra enrichment.

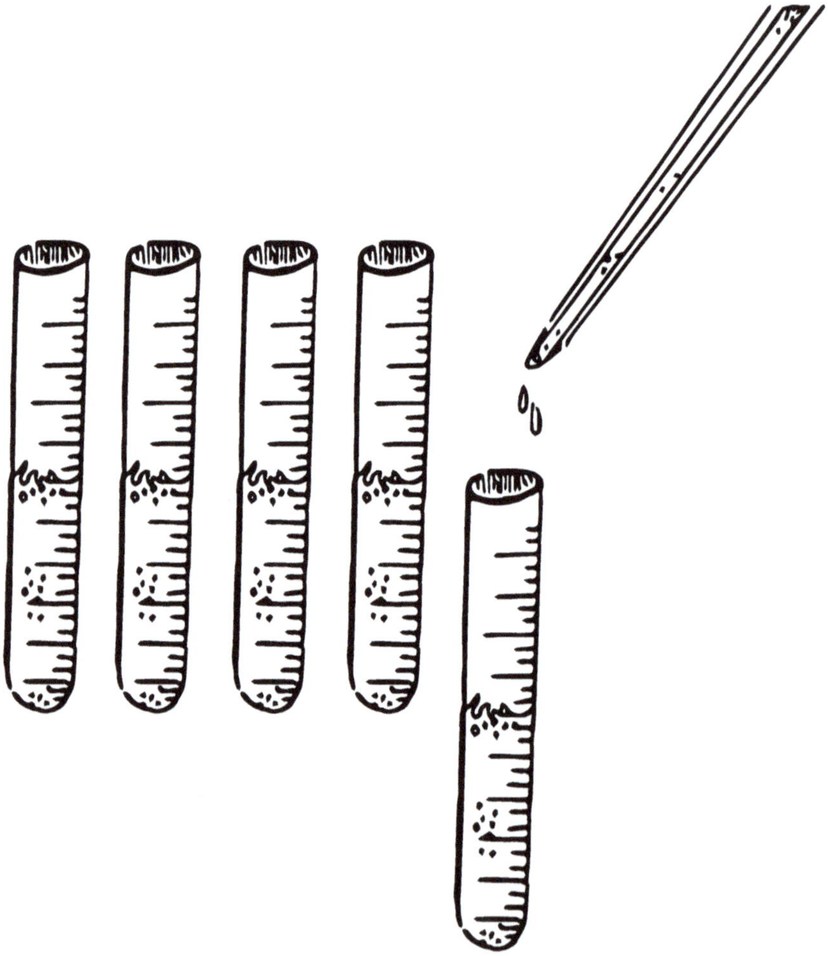

To measure objects within the field of microbiology, important algebraic skills such as ratio and unit conversion are needed. These skills are often applied in labs and science experiments. In this lesson, you will learn and practice the process of serial dilution and calculate the cell viability of a given sample.

DATA SHEET

PART 1 PROCEDURE: SERIAL DILUTIONS			
	Taste	**Concentration**	**Scale**
Stock Solution			ones
1st dilution			tenths
2nd dilution			thousandths (milli)
3rd dilution			ten thousandths
4th dilution			hundred thousandths
5th dilution			millionths (micro)
6th dilution			ten millionths
7th dilution			hundred millionths
8th dilution			billionths (nano)
9th dilution			ten billionths
10th dilution			hundred billionths

PART 2: CELL VIABILITY COUNTING			
	total count	**viable count**	**% viability**
Trial 1			
Trial 2			
Trial 3			
Trial 4			
Totals			

$$\frac{\square \text{ viable “cells”}}{\square \text{ total “cells”}} = \square$$

$$\square \times 100 = \square\ \%$$

Appendix A

Grab and Go Lessons

Grab and Go 1 Make Your Own Electrophoresis Chamber!

Introduction

DNA fragments can be separated by placing samples in a gel matrix and applying a low voltage current. Because DNA is negatively charged, the fragments move through the gel towards the positively charged end and are distributed according to size – smaller molecules travel faster than larger molecules, which are retained by the gel matrix. The distribution pattern of DNA fragments can then be used in many biotechnology applications, from the identification of members of a population of bacteria to the identification of suspects in forensics cases.

Battery setup

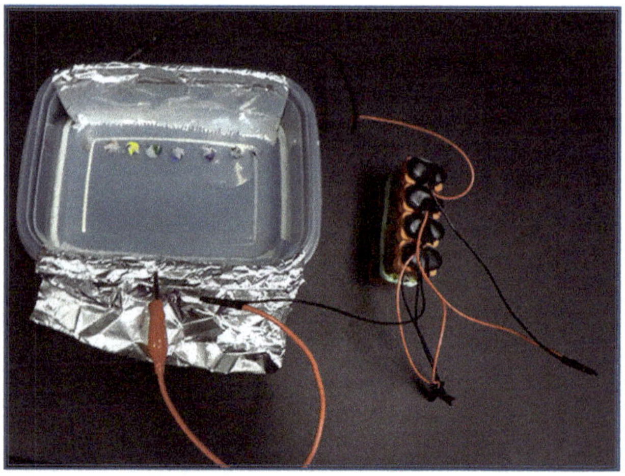

Beginning of experiment

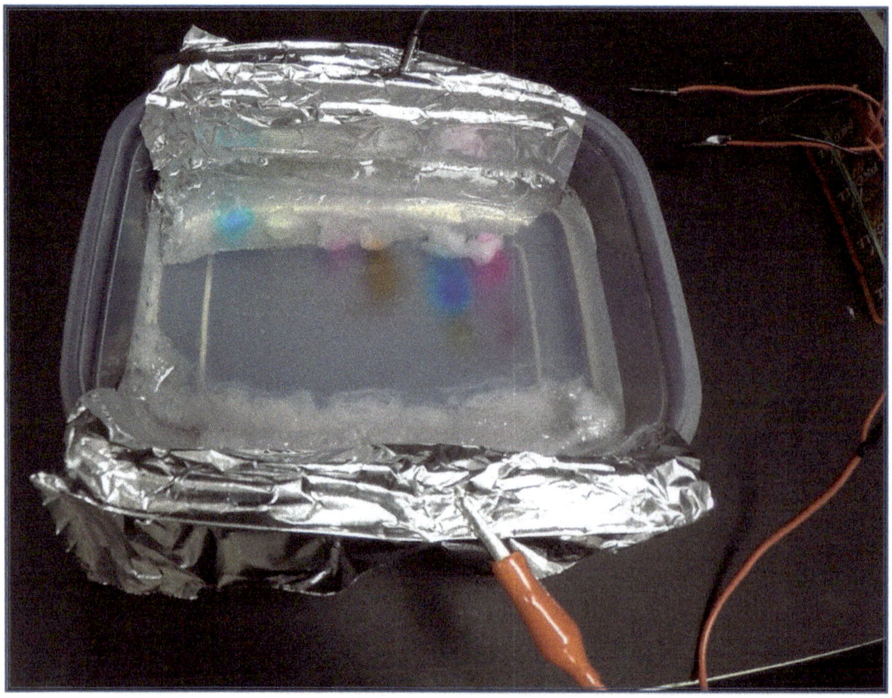

End of experiment results

Grab and Go 1 Make Your Own Electrophoresis Chamber!

Learner Outcomes

- Construct and run a gel electrophoresis to separate a mixture of colors.
- Recognize that this model represents how gel electrophoresis can separate different sizes of DNA fragments.

Skill Level: Advanced, ages 14-18

Success Indicators

- Participants will construct a working model of gel electrophoresis with separation of colored ink samples.

Life Skills

Marketable Skills

Time Needed

2-3 hours

Space

Kitchen or room with counter, microwave, and water source

Suggested Group Size

3-4 participants per group

Materials List

- Rectangular plastic container
- Aluminum foil
- Agar-Agar powder (from Asian food store or online supplier)
- Baking soda
- Water
- Water-soluble markers
- Measuring spoons
- Drinking straws - 4
- Tweezers
- Tissue – facial or bath
- 4- 9-V batteries
- 4- 9-V battery clip connectors
- 2 wires with alligator clips
- block connectors or electrical tape
- scissors or wire strippers
- modeling clay
- tape
- microwave
- 1-C measuring cup (2)
- Technology to take pictures or video of the experiment (optional-consider allowing participants to use personal devices for this step at your discretion)

Alternative: One gel electrophoresis may be created for the entire class if time or materials are unavailable for groups to create their own.

Important!!
Connect the batteries using the exact directions and materials suggested in this lesson. Using other methods of powering the gel electrophoresis may lead to battery damage or dangerous conditions.
Do not leave the running experiment unattended.

This grab and go can be used as a complement to Lesson 4: Knock, Knock, which explains DNA in greater detail and has participants model two forms of DNA sequencing. If this lesson will be taught independently, consider supplementing the information in this lesson with information from Lesson 4.

Grab and Go 1 ✱ Make Your Own Electrophoresis Chamber!

Before the Activity

- Assemble materials needed for making gel chamber according to instructions.
- The gel chamber can be constructed and the gel created before class if necessary up to 24 hours in advance.
- Prepare technology to present the Gel Electrophoresis web activity.
- Make copies of the Make your Own Gel Electrophoresis Apparatus! for each group (if each group will make their own).

Opening Question:

- How can pieces of DNA be separated from one another?

Let's Do It!

1. Ask the Opening Question.

2. Explain to the participants that you can't just pick up DNA pieces with tweezers! Guide participants through the web activity at the following link:

   http://learn.genetics.utah.edu/content/labs/gel/

3. Once participants have a grasp of the process of gel electrophoresis, introduce the Make your Own Gel Electrophoresis Apparatus! activity. Distribute the handout to each group.

4. Assist groups as they complete the steps for this learning experience as needed. Direct participants as they share resources (water, microwave access).

5. Optional: Participants may take pictures or videos of their gel electrophoresis tray as the experiment runs.

6. Ask the Share questions.

   - Did any gel chamber experiments not work? If so, why not?
   - Which color in the ink mixtures traveled farther in the gel? What does that tell you about the size of the color molecules?
   - Did all the samples of a color give the same results? Would you expect samples of DNA having the same sizes of fragments to yield similar patterns?

7. Ask the Reflect question.

   - How useful would it be to have the DNA fragment pattern information available for investigating disease outbreaks?

GRAB AND GO 1 MAKE YOUR OWN ELECTROPHORESIS CHAMBER!

8. Ask the Generalize question.
    - Can you think of some other instances where it would be an advantage to have DNA fragment patterns available for comparison?
9. Ask the Apply questions.
    - Which careers do you believe may use DNA separation technology (for example, forensics, paternity testing, or epidemiology)?
    - What kind of training might a person need to find employment in one of these fields?

Talk It Over

SHARE 1	REFLECT 2
Did any gel chamber experiments not work? If so, why not? Which color in the ink mixtures traveled farther in the gel? What does that tell you about the size of the color molecules? Did all the samples of a color give the same results? Would you expect samples of DNA having the same sizes of fragments to yield similar patterns?	How useful would it be to have the DNA fragment pattern information available for investigating disease outbreaks?
GENERALIZE 3	**APPLY 4**
Can you think of some other instances where it would be an advantage to have DNA fragment patterns available for comparison?	Which careers do you believe may use DNA separation technology (for example, forensics, paternity testing, or epidemiology)? What kind of training might a person need to find employment in one of these fields?

GRAB AND GO 1 MAKE YOUR OWN ELECTROPHORESIS CHAMBER!

Term and Concept Discovery

DNA pairing: bases on one strand are paired specifically with a base on the opposite strand – A always pairs with T and C with G

DNA sequences: each DNA sequence is unique to an individual organism and can be used for identification

gel electrophoresis: a technique for separating protein molecules of varying sizes in a mixture by moving them through a block of gel, as of agarose or polyacrylamide, by means of an electric field,with smaller molecules moving faster and therefore farther than larger ones.

Acknowledgments

"Kitchen Electrophoresis" Wellcome Trust Centre for Human Genetics

References

Homemade gel electrophoresis. https://bit.ly/3uPbhpd

LEARN MORE

Try your hand at a simulation of cutting DNA into pieces and separating them on a simulated gel in a real-world case study. See "Comparing DNA" from "Restoring the Chestnut" activity. https://bit.ly/34I53Nm

Education Standard(s):

NGSS HS-LS3.A Each chromosome consists of a single very long DNA molecule, and each gene on the chromosome is a particular segment of that DNA. The instructions for forming species' characteristics are carried in DNA. All cells in an organism have the same genetic content, but the genes used (expressed) by the cell may be regulated in different ways. Not all DNA codes for a protein; some segments of DNA are involved in regulatory or structural functions, and some have no as-yet known function

NGSS HS-LS4.A Genetic information provides evidence of evolution. DNA sequences vary among species, but there are many overlaps; in fact, the ongoing branching that produces multiple lines of descent can be inferred by comparing the DNA sequences of different organisms. Such information is also derivable from the similarities and differences in amino acid sequences and from anatomical and embryological evidence.

Tags

DNA, biotechnology, gel electrophoresis

Grab and Go 1 Print Materials

Facilitator Print Materials

None

Group Handouts

Make your own gel electrophoresis apparatus!

Youth Handouts

None

MAKE YOUR OWN ELECTROPHORESIS APPARATUS!

What you need:

- Rectangular plastic food container
- Aluminum foil
- Agar-agar powder
- Baking soda
- Water
- Water-soluble markers
- Measuring spoons
- Drinking straws
- Tweezers
- Tissue - facial or bath
- Four 9V batteries
- Two wires with alligator clips
- Modeling clay or masking tape
- Tape
- Microwave
- Measuring cup, graduated cylinder, or other measuring container that shows volume in milliliters (mL)

Step 1: Assemble the gel box

- Tear off two pieces of foil and fold them over the long sides of the container. Make sure they fit all the way to the bottom.

Step 2: Prepare the gel comb

- Cut the straws into sections that are an inch taller than the gel container. Reserve two straws to create a cross piece.
- Create a "comb" of straws. Cut two pieces of tape to match the width of the plastic food container. Lay one piece of tape with the adhesive side up on a flat surface. Place the straws perpendicular to the tape at regular distances from one another. Secure the straws by placing the second piece of tape across the straws. (see illustration)
- Tape two straws across the straw comb to act as a cross piece to hold the comb 1/4" above the bottom of the gel chamber. (see illustration)
- Plug the open ends of the straws with a small amount of modeling clay or cover them with masking tape.
- Test the comb by resting the cross-piece straw on the edge of the gel chamber. The bottom (covered) ends of the straws should be about one-quarter inch above the bottom of the gel chamber.
- Tape the comb about one-half inch from the long side of the container.

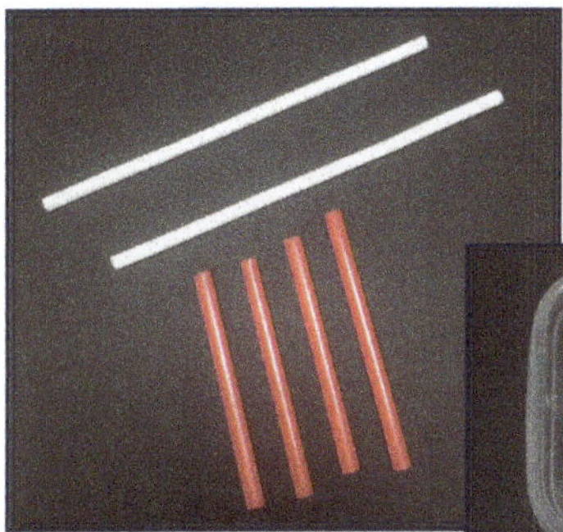
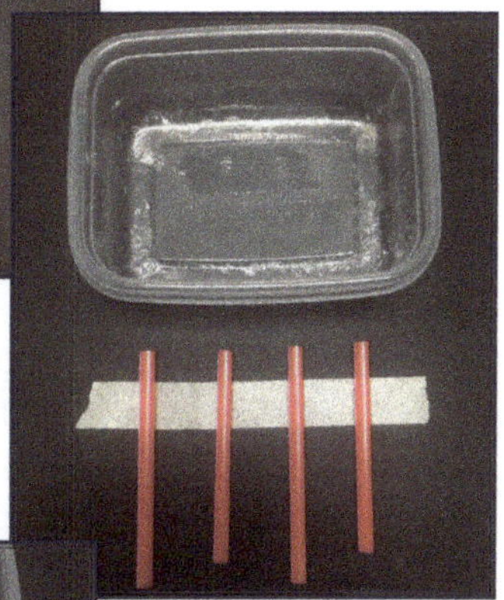
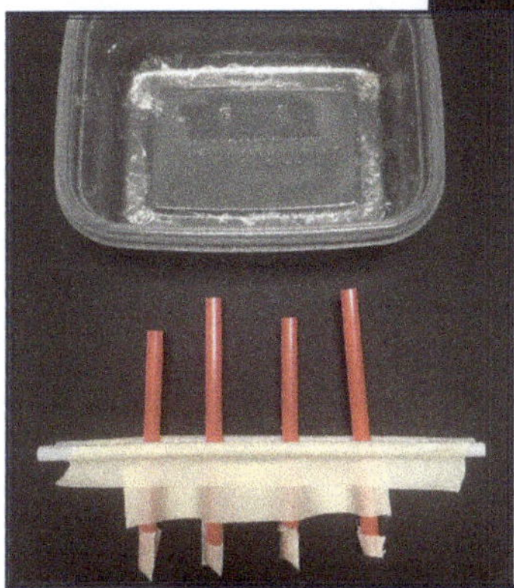
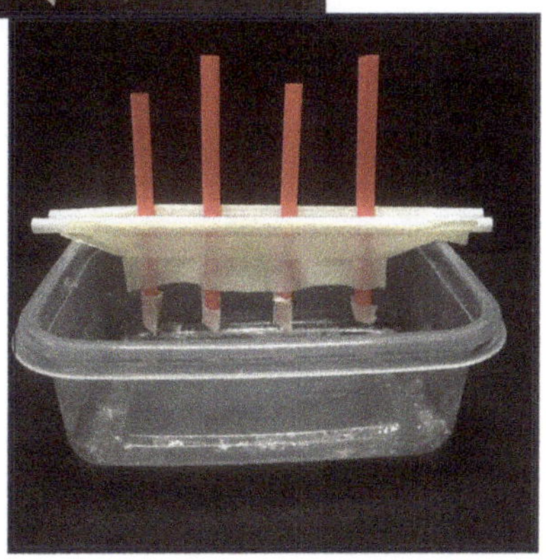

Step 3: Make the gel

- Measure ½ teaspoon of the agar-agar powder into a measuring cup. Add 200 mL water and stir.
- Microwave on high for 2 minutes. Stir to dissolve. If necessary, microwave up to 1 minute more. Avoid boiling over.
- Set gel aside. In a separate measuring cup, dissolve 1 teaspoon of baking soda in 250 mL water.
- Add 1 teaspoon of the baking soda mixture to the hot agar-agar and stir. Carefully add to the gel chamber and allow to set (about 20 minutes).
- Reserve the baking soda and water solution for later use.
- After the gel has set, carefully remove the straws. Small holes should remain in the gel. This is where you will load the samples.

Step 4: Prepare the power supply

- Connect the four 9V batteries with the clip connectors as shown in the picture.
- Connect the red alligator clip to the free red wire and the black alligator clip to the free black wire. Do not connect to the gel apparatus yet.
- Carefully strip the end off of the wires coming from the clip connectors. Twist the ends of the wires together as shown and wrap the exposed wires in electrical tape.

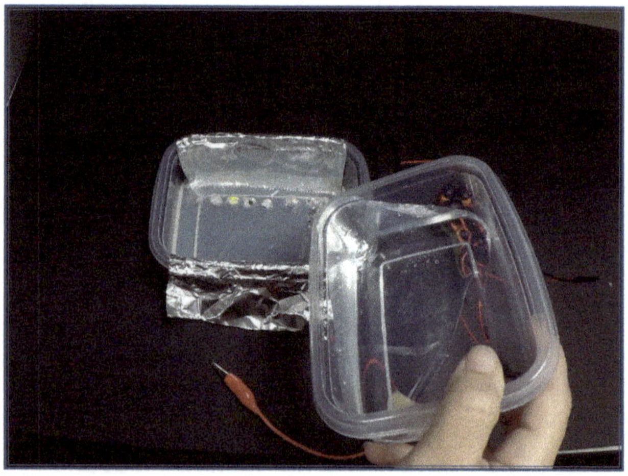

Step 5: Prepare the samples

- With a water-based marker, make a small spot of color on a piece of tissue. A spot ¼ inch in diameter is plenty.
- Tear out the spot and wad it up into a small ball.
- Repeat with markers of three different colors.
- With tweezers, push a ball of tissue into each of the wells.

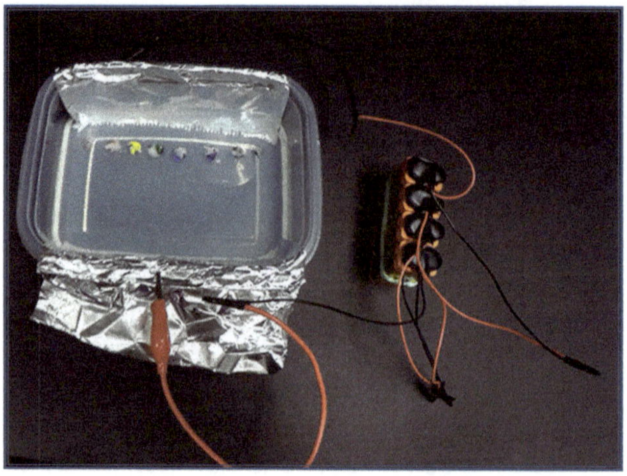

MAKE YOUR OWN ELECTROPHORESIS APPARATUS! CONTINUED

Step 6: Run the gel

- Measure 200 mL of water. Add 1 tsp of the baking soda mixture from Step 3 to the water and stir.
- Carefully pour this solution onto the gel away from the wells. Pour until about ½ inch of solution covers the gel.
- Connect the red (positive) alligator clip to the side of the gel apparatus opposite from the wells.
- Connect the black (negative) alligator clip to the other side.
- Allow the gel to run for about one hour or as long as you wish. You should see the color begin to move away from the wells after a few minutes.

Sample results:

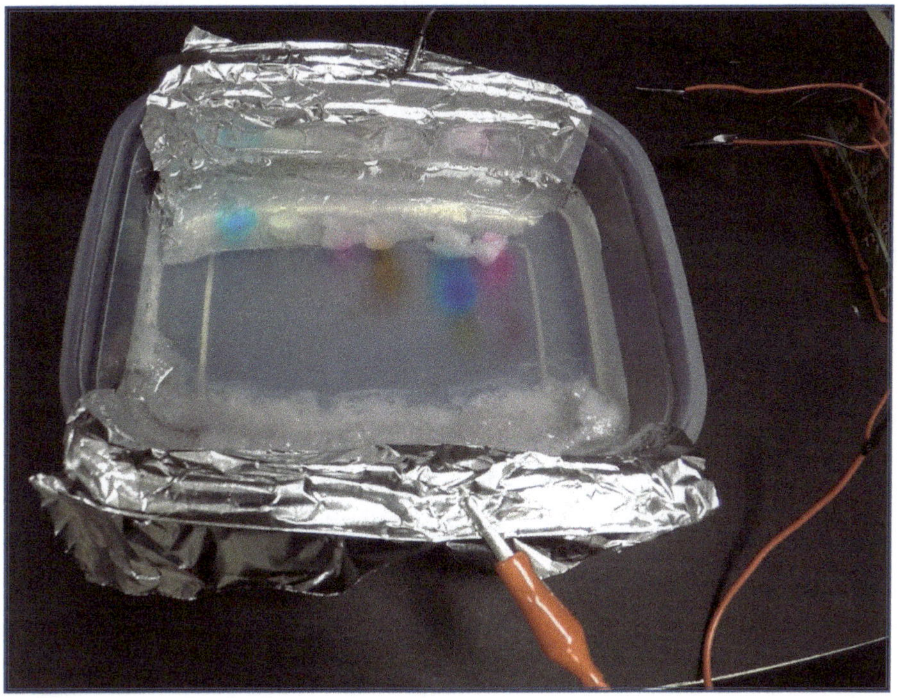

What's Happening?

Electrophoresis is the movement of charged molecules through a matrix due to the presence of an electrical voltage. Negatively charged molecules such as dyes and DNA move towards the positive pole. The molecules move at different speeds due to their sizes - small molecules can move faster through the gel than large molecules can. Therefore, a mixture containing different sizes of molecules can be separated.

Grab and Go 2 ✱ Plate 'Em Up

Introduction

In any given room in the house, there are millions of microorganisms that are unseen by human occupants. Most of these never get enough competitive advantage to make a colony that is big enough to see or are non-colony forming. This activity is designed to show a gross method of identifying bacteria and other microorganisms by the shapes they make as they grow.

Learner Outcomes

- View how bacterial colonies form.

Skill Level: Advanced, ages 14-18

Success Indicators

- Participants will identify types of colonial growth.

Life Skills

Critical thinking, Decision Making

Time Needed

30-60 minutes

Space

2'x4' countertop

Suggested Group Size

2-3 participants

Materials List

- 2 petri dishes
- Magnifying glass or microscope
- Sponge
- Safety gloves
- Scissors
- Permanent marker
- Technology to research and classify bacterial colonies

Before the Activity

- Prepare materials for each group at stations.

Opening Question:

- What do you think is living on your sponge?

Let's Do It!

1. Direct participants to join their small groups and observe the materials at their station.

2. Ask the Opening Question.

3. Direct participants to complete the following steps:

    - Use the permanent marker to label the petri dishes #1 and #2.
    - Put on the safety gloves.
    - Cut a small piece of the sponge before use and rub it on petri dish #1. Close the petri dish and set it aside.
    - Wet the sponge with water and pick a surface on which to rub the sponge for 3 minutes.
    - Squeeze as much water as you can out of the sponge.
    - Rub one surface of the sponge on petri dish #2 and set it aside
    - Predict what you will find on the second plate and place the list under the plate.

4. Two days later, compare the list of what you thought would grow to what is growing on the dish.

Grab and Go 2 Plate 'Em Up

5. Ask the Share questions.

    - What grew on your plate?
    - What was needed for the bacteria to grow on the plate?

6. Ask the Reflect question.

    - Why did your plate differ from another group's plate?

7. Ask the Generalize question.

    - Can you make a statement about what the microorganisms need to live?

8. Assign the Apply topic.

    - Research the shape of the colony and find possible microbes that make that shape with their colonies. What role do these microbes play in the environment?

9. Using these links, classify any colonies that have grown on the dishes.

Microbiology Society - https://bit.ly/3z023d9

Biology Libre Texts - https://bit.ly/3fOjSEA

Baker College Research Guides - https://bit.ly/3vOMLpF

10. *Note: If you have pre-designated locations, you can make the plates two days ahead of time and show participants the results on the day of the learning experience. Later participants can compare and contrast their plates with the sample plates.

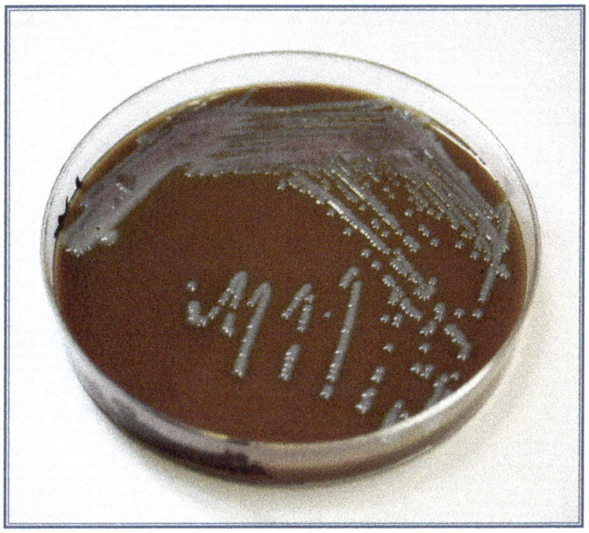

Grab and Go 2 ✱ Plate 'Em Up

Talk It Over

Share 1	Reflect 2
What grew on your plate? What was needed for the bacteria grow on the plate?	Why did your plate differ from another group's plate?
Generalize 3	**Apply 4**
Can you make a statement about what the microorganisms need to live?	Research the shape of the colony and find possible microbes that make that shape with their colonies. What role do these microbes play in the environment?

Variations

Divide the sponge into pieces and let every group member pick a different surface. Divide the second petri dish into 2-3 sections with a marker and rub each sponge on one section.

Term and Concept Discovery

Agar: nutritious media for microbes to grow on

Petri Dish: common container to hold agar

Colony: collection of the same type of microbes that grow from a single cell

Morphology: study of the form and structure of things

Education Standard(s):

HS-LS2-3. Construct and revise an explanation based on evidence for the cycling of matter and flow of energy in aerobic and anaerobic conditions.
HS-LS1-7. Use a model to illustrate that cellular respiration is a chemical process whereby the bonds of food molecules and oxygen molecules are broken and the bonds in new compounds are formed resulting in a net transfer of energy.

Tags

Populations, cell division, microbiology

Grab and Go 3 • Stay Away From Our Food, You Germs!

Introduction

The Centers for Disease Control (CDC) estimates that each year roughly 1 in 6 Americans (or 48 million people) gets sick, 128,000 are hospitalized, and 3,000 die from foodborne diseases. What is the cause of this? Harmful bacteria, viruses, and microbes are causing most of the illnesses in the United States. This estimate shows that there is still much work to be done—specifically in focusing efforts on food safety. Creating food safety awareness in the community is important.

"The Partnership for Food Safety Education" is a non-profit organization. Its mission is to end illness and death from foodborne infection in the United States. One of its campaigns is the "Fight BAC" campaign. BAC is a term for bacteria. The campaign gives four important steps in keeping our food safe. These four steps are Clean, Separate, Cook, and Chill. More information can be found on the following website:
http://www.fightbac.org/safe-food-handling

In this community project/activity, students will be communicating with various community members who take care of our food supply. Then they will be creating a poster that summarizes the core four practices of food safety- Clean, Separate, Cook, and Chill.

Grab and Go 3 ✱ Stay Away from Our Food, You Germs!

Learner Outcomes

- Describe the importance of consuming safe food.
- Explain general guidelines that should be followed in the safe handling of all foods.
- Create and present a poster as a public service advertisement for food safety which will feature a specific method for preventing foodborne illness.

Skill Level: Intermediate

Success Indicators

- Participants will discuss and explain what it means for our food to be safe.
- Participants will discuss and explain where, when, and how our food supply is vulnerable to contamination.
- Participants will discuss and explain the effects of foodborne contamination on human health.

Life Skills

Sharing, Nurturing Relationships, Social skills, Communication, Keeping Records

Time Needed

Two sessions of 90 minutes each

Space

Table and chairs

Suggested Group Size

2 students per group

Materials List

- Web enabled computers
- Poster board and markers or technology to prepare electronic presentations
- Fight BAC! color brochure (1 per participant)

fightbac.org

Grab and Go 3 ✱ Stay Away from Our Food, You Germs!

Before the Activity

- Consider providing parents with information regarding the public contact portion of this lesson before assigning participants an activity.
- Make copies of the Fight BAC! color brochure (1 per participant).
- Prepare technology to show the Food Safety Introduction presentation and the four YouTube videos referenced in the presentation.

Part One

Opening Question:

- What is wrong with these pictures?

Let's Do It!

1. Show the Food Safety Introduction presentation. ** Note: On slides 14-17, the graphic is a link to a short video that can be accessed if your computer is connected to the Internet.

2. After the presentation, give each student a "Fight BAC!" brochure. Explain to the participants that they are officially BAC Fighters and are responsible for creating awareness about food safety.

3. Divide the students into groups of 2-3 students. Each group will perform two tasks.

4. Assign each group **one assignment** from the following list. Each assignment has two tasks.

    - The **first task** is real-life community interaction.
    - The **second task** is creating a poster or other presentation as a public service advertisement of the food safety process.

5. Allow participants to conduct research using web enabled computers or other devices to identify potential interviewees. Participants may begin to contact interviewees by email, phone, or other methods during this portion of the lesson. Direct and assist participants as necessary.

Grab and Go 3 ✹ Stay Away from Our Food, You Germs!

Assignment	First Task	Second Task
1.	Interview the manager of a grocery store to find out what food safety guidelines they follow. Ask if the manager has heard about the "Fight BAC!" campaign. How does the store follow the four steps? With permission, record the temperature at various places in the store.	Create a poster to highlight how the grocery store uses the four "Fight BAC!" steps.
2.	Visit a food processing plant and observe the food products being processed. List precautions taken to assure safety and a quality finished product. Ask if the manager has heard about the "Fight BAC!" campaign. How does the plant follow the four steps?	Create a poster to highlight how the food processing plant uses the four "Fight BAC!" steps.
3.	Interview a food inspector. Ask the inspector to explain his or her job duties, common food safety violations often encountered, the importance of a safe and secure food source, and general guidelines that the average person should practice. Ask if the inspector promotes the "Fight BAC!" campaign. How does he or she follow the four steps?	Create a poster to highlight the four "Fight BAC" steps checked in food safety violations.
4.	Visit a school cafeteria to learn the safety precautions that must be followed in storing, processing, preparing, and serving food at school. Ask the food service workers if they have heard about the "Fight BAC!" campaign. How do they follow the four steps?	Create a poster to highlight the four "Fight BAC!" steps in the school cafeteria.

Grab and Go 3 ✱ Stay Away from Our Food, You Germs!

Part Two

1. Once the participants have completed an interview and a project, facilitate a discussion among the participants before the groups present their findings.

2. Ask the Reflect questions.

   - What does it mean for our food to be "safe"? How do we know when it is safe?
   - Where, when, and how is our food supply vulnerable to contamination?
   - What are the effects of foodborne contamination on our health?
   - What can be done to prevent contamination?

3. Ask the Generalize question.

   - Whose responsibility is it to keep our food supply safe?

4. Once the whole class discussion has concluded, give time for the groups to present to the class as a whole.

Talk It Over

Share 1	Reflect 2
Participants display and present their poster in class to explain and discuss various food safety guidelines.	What does it mean for our food to be "safe"? How do we know when it is safe? Where, when, and how is our food supply vulnerable to contamination? What are the effects of foodborne contamination on our health? What can be done to prevent contamination?
Generalize 3	**Apply 4**
Whose responsibility is it to keep our food supply safe?	Participants will apply their learning through their poster or other format and effective communication during the presentation.

GRAB AND GO 3 ✲ STAY AWAY FROM OUR FOOD, YOU GERMS!

Term and Concept Discovery

Binary fission: the proces of a bacterium splitting and becoming two

Contamination: the act of contaminating, or of making something impure or unsuitable by contact with something unclean, bad, etc.

Cross contamination: the passing of bacteria, microorganisms, or other harmful substances indirectly from one patient to another through improper or unsterile equipment, procedures, or products

Foodborne illness: food poisoning

Chill: keep food at or below 40 degrees Fahrenheit or frozen until ready to cook
Defrost: to thaw or partially thaw

Thaw: bring to a more normal temperature, especially to room temperature

References

http://www.fightbac.org/safe-food-handling

Education Standard(s):

Bio.2.1.4 Explain how disease can disrupt ecosystem balance. (Examples: AIDS, influenza, tuberculosis, Dutch Elm Disease, Pfiesteria, etc.)
Bio.2.2.1 Infer how human activities (including population growth, pollution, global warming, burning of fossil fuels, habitat destruction and introduction of nonnative species) may impact the environment.
Bio.3.4.3 Explain how various disease agents (bacteria, viruses, chemicals) can influence natural selection.

Next generation science Standards:

HS-LS1-3 Plan and conduct an investigation individually and collaboratively to produce data to serve as the basis for evidence, and in the design: decide on types, how much, and accuracy of data needed to produce reliable measurements and consider limitations on the precision of the data (e.g., number of trials, cost, risk, time), and refine the design accordingly.
HS-LS1-3 Scientific inquiry is characterized by a common set of values that include: logical thinking, precision, open-mindedness, objectivity, skepticism, replicability of results, and honest and ethical reporting of findings.
HS-LS2-2, HS-LS2-6 A complex set of interactions within an ecosystem can keep its numbers and types of organisms relatively constant over long periods of time under stable conditions. If a modest biological or physical disturbance to an ecosystem occurs, it may return to its more or less original status (i.e., the ecosystem is resilient), as opposed to becoming a very different ecosystem. Extreme fluctuations in conditions or the size of any population, however, can challenge the functioning of ecosystems in terms of resources and habitat availability.

Tags

cross contamination

Grab and Go 3 Print Materials

Facilitator Print Materials
None

Group Handouts
None

Youth Handouts
Fight BAC! color brochure

Be a BAC Fighter

Make the meals and snacks from your kitchen as safe as possible. **CLEAN:** wash hands and surfaces often; **SEPARATE:** don't cross-contaminate; **COOK:** to proper temperatures, and **CHILL:** refrigerate promptly. Be a BAC Fighter and *Fight BAC!*®

For More Information about Safe Food Handling and Preparation

USDA's Meat and Poultry Hotline
1-888-MPHotline (1-888-674-6854);
TTY 1-800-256-7072

www.foodsafety.gov

To locate food safety educators in your state or community:

Cooperative Extension Service, local offices:
www.csrees.usda.gov/Extension/index.html

Food and Drug Administration Public Affairs Specialists:
www.fda.gov/ora/fed_state/dfsr_activities/dfsr_pas.html

See *www.fightbac.org* for free brochures, fact sheets, stickers, and other great stuff!

Sign up to be a BACFighter at www.fightbac.org

Apply the heat... and Fight BAC!®

Cooking food to the proper temperature kills harmful bacteria. So *Fight BAC!*® by thoroughly cooking your food as follows:

SAFE COOKING TEMPERATURES
as measured with a food thermometer

	Internal temperature
Ground Meat & Meat Mixtures	
Beef, Pork, Veal, Lamb	160°F
Turkey, Chicken	165°F
Fresh Beef, Veal, Lamb	
Medium Rare	145°F
Medium	160°F
Well Done	170°F
Poultry	
Chicken & Turkey, whole	165°F
Poultry parts	165°F
Duck & Goose	165°F
Stuffing (cooked alone or in bird)	165°F
Fresh Pork	
Medium	160°F
Well Done	170°F
Ham	
Fresh (raw)	160°F
Pre-cooked (to reheat)	140°F
Eggs & Egg Dishes	
Eggs	Cook until yolk & white are firm
Egg dishes	160°F
Seafood	
Fin Fish	145°F or flesh is opaque & separates easily with fork
Shrimp, Lobster & Crabs	flesh pearly & opaque
Clams, Oysters & Mussels	Shells open during cooking
Scallops	milky white or opaque & firm
Leftovers & Casseroles	165°F

2005 Partnership for Food Safety Education

FIGHT BAC!®

FIGHT FOODBORNE BACTERIA

Four Simple
Steps to
Food Safety

www.fightbac.org

BAC (foodborne bacteria) could make you and those you care about sick. In fact, even though you can't see BAC—or smell him, or feel him—he and millions more like him may have already invaded the food you eat. But you have the power to *Fight BAC!*®.

Foodborne illness can strike anyone. Some people are at a higher risk for developing foodborne illness, including pregnant women, young children, older adults and people with weakened immune systems. For these people the following four simple steps are critically important:

CLEAN: *Wash hands and surfaces often*

Bacteria can be spread throughout the kitchen and get onto hands, cutting boards, utensils, counter tops and food. To *Fight BAC!*®, always:

- Wash your hands with warm water and soap for at least 20 seconds before and after handling food and after using the bathroom, changing diapers and handling pets.
- Wash your cutting boards, dishes, utensils and counter tops with hot soapy water after preparing each food item and before you go on to the next food.
- Consider using paper towels to clean up kitchen surfaces. If you use cloth towels wash them often in the hot cycle of your washing machine.
- Rinse fresh fruits and vegetables under running tap water, including those with skins and rinds that are not eaten.
- Rub firm-skin fruits and vegetables under running tap water or scrub with a clean vegetable brush while rinsing with running tap water.

SEPARATE: *Don't cross-contaminate*

Cross-contamination is how bacteria can be spread. When handling raw meat, poultry, seafood and eggs, keep these foods and their juices away from ready-to-eat foods. Always start with a clean scene— wash hands with warm water and soap. Wash cutting boards, dishes, countertops and utensils with hot soapy water.

- Separate raw meat, poultry, seafood and eggs from other foods in your grocery shopping cart, grocery bags and in your refrigerator.
- Use one cutting board for fresh produce and a separate one for raw meat, poultry and seafood.
- Never place cooked food on a plate that previously held raw meat, poultry, seafood or eggs.

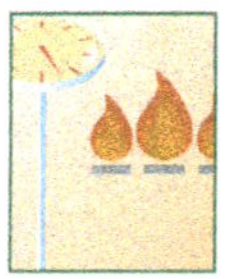

COOK: *Cook to proper temperatures*

Food is safely cooked when it reaches a high enough internal temperature to kill the harmful bacteria that cause illness. Refer to the chart on the back of this brochure for the proper internal temperatures.

- Use a food thermometer to measure the internal temperature of cooked foods. Make sure that meat, poultry, egg dishes, casseroles and other foods are cooked to the internal temperature shown in the chart on the back of this brochure.
- Cook ground meat or ground poultry until it reaches a safe internal temperature. Color is not a reliable indicator of doneness.
- Cook eggs until the yolk and white are firm. Only use recipes in which eggs are cooked or heated thoroughly.
- When cooking in a microwave oven, cover food, stir and rotate for even cooking. Food is done when it reaches the internal temperature shown on the back of this brochure.
- Bring sauces, soups and gravy to a boil when reheating.

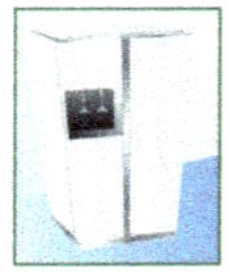

CHILL: *Refrigerate promptly*

Refrigerate foods quickly because cold temperatures slow the growth of harmful bacteria. Do not over-stuff the refrigerator. Cold air must circulate to help keep food safe. Keeping a constant refrigerator temperature of 40°F or below is one of the most effective ways to reduce the risk of foodborne illness. Use an appliance thermometer to be sure the temperature is consistently 40°F or below. The freezer temperature should be 0°F or below.

- Refrigerate or freeze meat, poultry, eggs and other perishables as soon as you get them home from the store.
- Never let raw meat, poultry, eggs, cooked food or cut fresh fruits or vegetables sit at room temperature more than two hours before putting them in the refrigerator or freezer (one hour when the temperature is above 90°F).
- Never defrost food at room temperature. Food must be kept at a safe temperature during thawing. There are three safe ways to defrost food: in the refrigerator, in cold water, and in the microwave. Food thawed in cold water or in the microwave should be cooked immediately.
- Always marinate food in the refrigerator.
- Divide large amounts of leftovers into shallow containers for quicker cooling in the refrigerator.
- Use or discard refrigerated food on a regular basis. Check USDA cold storage information at ***www.fightbac.org*** for optimum storage times.

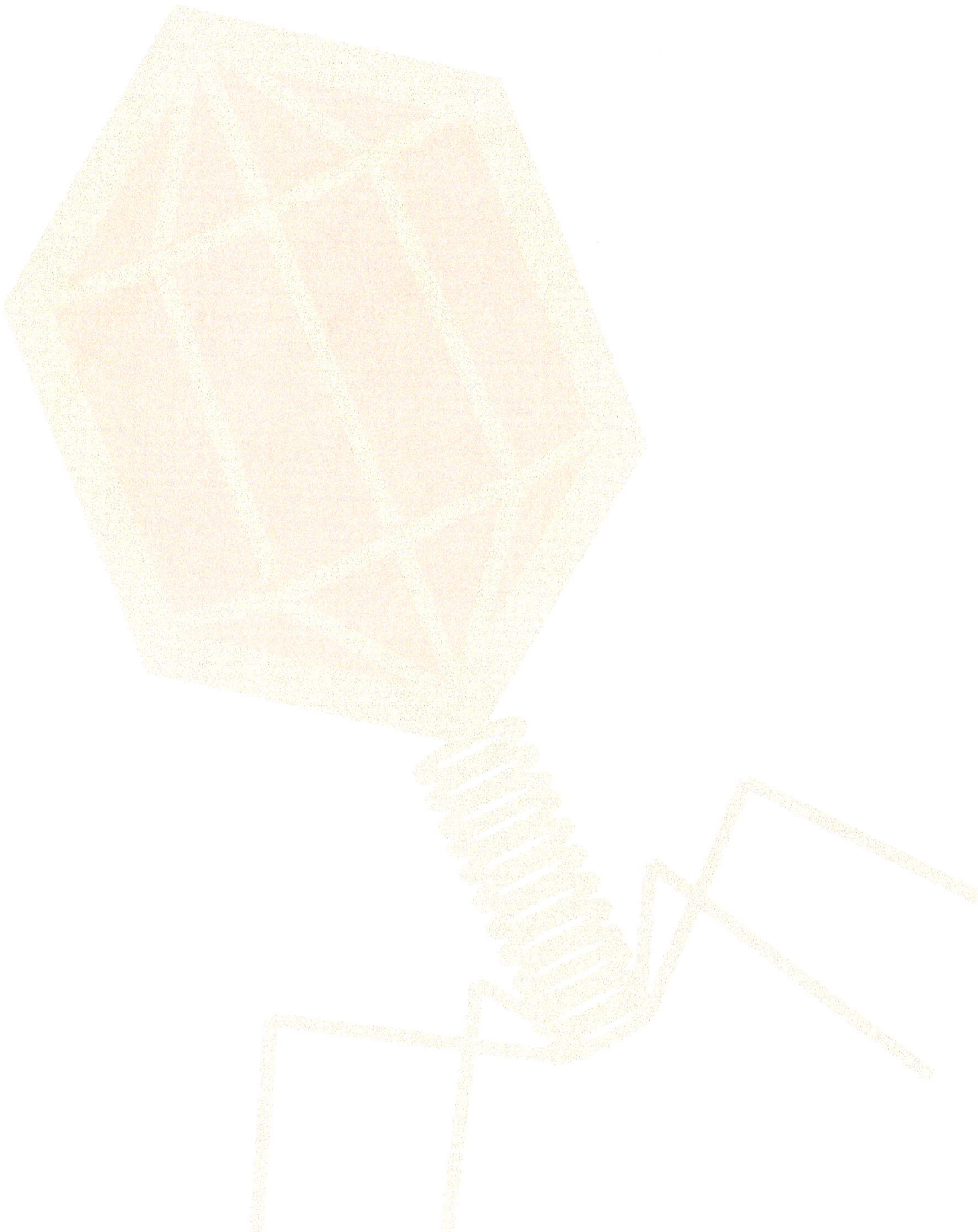

Appendix B

Materials List

Materials List for High School Group

1 group=4 youth

Lesson 1		
Consumables	Non Consumables	Items to be printed
Poster paper or clean 8 ½" x 11" paper	"Body Bacteria: Exploring the Skin's Microbial Metropolis" article https://bit.ly/3piEFTD	Fortune tellers
Markers, scissors		Set of 12 Bacterial Mug Shots
		Bacteria Dichotomous Key

Lesson 2A		
Consumables	Non Consumables	Items to be printed
3 green bananas (per group)	Small bowl or plate (1 per group)	Dare to Infect Me Introduction and Lab handout (1 per participant)
1 rotten banana (per group)	Knife (1 per group)	
Spoon (one or more per group)	Antibacterial soap	
Craft sticks (several per group)	Permanent marker (1 per group)	
3 sealable plastic bags lettered A-C with a permanent marker (per group)		
Paper towels		
Petroleum jelly		

Lesson 2B		
Consumables	Non Consumables	Items to be printed
1 braided candy rope (1" to 2" segment per group)	Scissors (1 pair per group)	Code Red Introduction and Lab handout (1 per participant)
Gummy bears (minimum 24 per group)		
Round toothpicks (minimum 12 per group)		
Gummy rings (2 per group)		
Juice-filled fruit snacks (6 per group)		
Large round cookies (1 per group)		
Paper towels		

Materials List for High School Group (continued)

1 group=4 youth

Lesson 3		
Consumables	Non Consumables	Items to be printed
1 prepared note card per participant	Web enabled computers (1 piece of technology per group)	1 Design a Disease Museum assignment sheet per pair of participants
Tri fold poster board		

Lesson 4		
Consumables	Non Consumables	Items to be printed
3 ziplock bags (large) (for 3 groups)	Test tube (1 per participant)	"Unknown" DNA sequences – as many as desired. 1 unique sequence per "bead" – each sequence has a "bar code" sequence at the beginning of the unknown sequence.
33 pipecleaners (for 3 groups)	Dropper (1 per group)	Nucleotide labels – "A" "T" "C" "G" – can have multiple sets
colored beads (4 colors)	Markers	"Bar code" sequences that match "unknown"
Bathroom cup (3-4 oz) (1 per participant)		
Dish detergent (1 large container)		
Sports Drink (ex. Gatorade) (enough for each participants to have 3-4 oz.)		
Meat Tenderizer (Ex. Adolph's) (1 container)		
90% Isopropyl (rubbing) alcohol – COLD (from the freezer – take out just before use) (1 large container)		
Safety gloves (1 pair per participant)		
Chart paper or whiteboard for Reporters		

Materials List for High School Group (continued)

1 group=4 youth

Lesson 5		
Consumables	Non Consumables	Items to be printed
	Five differently colored pens or pencils or markers per group	One or more game board(s) per group
		Scenario Sheet (1 per group)
		Event Cards (1 sheet per group)
		Activity Data Sheets (1 per group per scenario)

Lesson 5 Note:
Participants will play this game through four or five different scenarios. Please choose one of the following options for game play and prepare your materials accordingly:

- Dry erase - Participants will make dots or tally marks on a dry erase board for the bacteria that are added or subtracted in each scenario.
    - Materials needed: one dry erase board per group and an assortment of dry erase markers

- Laminated game board: Participants will use dry erase markers to make dots or tally marks directly on the laminated game board for the bacteria that are added or subtracted in each scenario.
    - Materials needed: one laminated game board per group and an assortment of dry erase markers

- Multiple game boards: Participants will use a different game board for each scenario of the game and make marks directly on the game board for the bacteria that are added or subtracted in each scenario.
    - Materials needed: one printed game board PER SCENARIO, per group (4-5 game boards per group)

Lesson 6		
Consumables	Non Consumables	Items to be printed
Red Drink Mix (premixed drink mix) (1 large container)	100 mL Graduated cylinder (one per group)	Data Sheet (1 per participant)
Clear plastic cups (12 per group)	Pitcher (for mixing solutions)	
"Fun Size" M & Ms packages (4 per group)	Digital scale	

Materials List for High School Group (continued)

1 group=4 students

Grab and Go 1		
Consumables	Non Consumables	Items to be printed
Rectangular plastic container (1 per chamber that will be created)	Water-soluble markers (1 set of 5-8 per chamber)	Make your Own Gel Electrophoresis Apparatus! handout (1 per chamber)
Aluminum foil (1 large roll)	Measuring spoons	
Agar-Agar powder (from Asian food store or online supplier)	Tweezers (1 pair per chamber)	
Baking soda (1 box)	9-V battery clip connectors (4 per chamber)	
Water	Wires with alligator clips (2 per chamber)	
Drinking straws (4 per chamber)	Block connectors or electrical tape	
Tissue – facial or bath (1 roll)	Scissors or wire strippers (1 per chamber)	
9-V batteries (4 per chamber)	Microwave	
Modeling clay	1-C measuring cup (2 per chamber)	
Tape	Technology to take pictures or video of the experiment (optional)	

Grab and Go 2		
Consumables	Non Consumables	Items to be printed
Sponge	2 petri dishes	
Safety gloves	Magnifying glass or microscope	
	Scissors	
	Permanent marker	

Grab and Go 2		
Consumables	Non Consumables	Items to be printed
Poster board and markers or technology to prepare electronic presentations	Web enabled computers	Food Safety Quiz! (1 per participant)
		Fight BAC! color brochure (1 per participant)

Appendix C

Teaching tips and techniques

Appendix C: Teaching Tips and Techniques

In this section, you will find information about some of the terms and structures used in designing the actual lessons. For information about the scientific vocabulary, please see the glossary.

The decision has been made to use the term "learners" instead of "youth" in this section, as these techniques work with adults as well.

Term	Anticipation Guides (also sometimes called Anticipation Sets)
What it is	Anticipation Guides are questions or checklists that learners complete prior to viewing or reading informational text.
How it works	Learners use their background knowledge to complete the Anticipation Guide prior to learning the information. This helps them to know what concepts they will encounter. It also helps learners to actively seek out and pay attention to the important ideas and major underlying concepts. After viewing or reading the informational text, learners go back and complete the same questions again. They are able to then compare their old knowledge with the knowledge gained as a result of the text.
When to use it	Use Anticipation Guides before and after viewing videos or reading informational text. Anticipation Guides are especially helpful for language learners who might otherwise be overwhelmed by the amount of information in a text or video.

Term	Claim-Evidence-Reasoning
What it is	Claim-Evidence-Reasoning is a lesson format that is based on scientific writing, argumentation, and thinking.
How it works	Learners encounter a research question and are asked to take a stand on a possible answer. At this point, the claim functions partly like a prediction and partly like a hypothesis, to use the language of the traditional scientific method. Learners then set about gathering evidence, which may include evidence from their personal experiences, evidence from printed or digital materials, and/or evidence from hands-on investigations. Once evidence has been gathered, students re-evaluate their initial claim. They have to decide which pieces of evidence are strongest and why. They also need to decide whether that evidence supports their initial claim, or if their initial claim needs to be revised.
When to use it	Use Claim-Evidence-Reasoning to support learners in developing scientific language and argument. Because the language mirrors the language in the Common Core State Standards, it can be used to support lessons in a variety of disciplines that require learners to identify and use evidence to support their thinking.

Term	Fishbowl Seminar
What it is	An alternative to a whole class discussion, the Fishbowl Seminar allows small group interaction and among students while involving the entire class.
How it works	The class creates two circles, an inner circle and an outer circle. The inner circle should have 6-8 seats or fewer. The facilitator asks open-ended, higher-order thinking questions. Participants on the outer circle take notes on the participants speaking in the inner circle. They are not allowed to speak or answer any questions. Participants in the inner circle discuss openly about the higher-order thinking questions that the facilitator poses. Participants in the outer circle can tap the shoulder of a participant in the inside circle when they have a strong point to make. This is how the circles change and how everyone gets incorporated into the discussion. Circle sizes can be changed to meet the needs of the group, but make sure to have more people in the outer circle than the inner circle. It is recommended to have at least 3-4 participants in the inner circle.
When to use it	The Fishbowl Seminar can be used as a class discussion tool when participant evaluation is important or to better hear each individual participant's input in a large group discussion.

Term	Four Corners
What it is	Four Corners is a formative technique for sharing out answers to a selected-response (multiple choice) question.
How it works	Learners are given a question along with several potential answers. The answers are displayed around the room, often in the corners. Learners move to the corner that matches their preferred response. Learners are allowed to share their reasons for their choice. They are also allowed to change their minds and move to a new location. A variation of Four Corners has students line up in front of their preferred choice. This creates a "human graph" which can be used to provide class data about misconceptions.
When to use it	Use Four Corners to determine what learners are thinking about key concepts and ideas in a quick, movement-based format.

Term	Formative Assessment Probes
What it is	A formative assessment probe is a question posed to gain an understanding of where learners are in their thinking. Frequently, assessment probes feature possible responses that mirror common thinking patterns among learners.
How it works	Formative Assessment Probes can be checklists, selected response questions, sorts, or other tasks that require students to articulate and apply scientific concepts. They are not used for grades; rather, they provide the base for the direction of the instruction. Learners are encouraged to explain their thinking as they complete the tasks.
When to use it	Use Formative Assessment Probes throughout a unit cycle to track learners' understanding of key concepts. Probes are especially useful at the beginning of a lesson to uncover existing misconceptions and prepare learners for the activity. The same probe can then be used at the close of the lesson to determine if learners' thinking changed or grew as a result of their work.

Term	Frayer Model
What it is	Frayer models are graphic organizer tools for developing vocabulary. It looks like this: 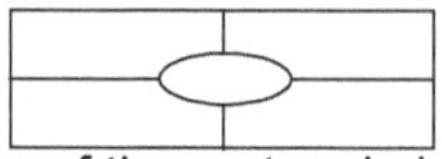 In the center of the rectangle is an oval. The new vocabulary word is written in the oval. Around the oval, the rectangular sections are labeled "Definition," "Characteristics," "Examples," and "Non-Examples." Occasionally, "Picture" or "Facts" is used in place of "Characteristics."
How it works	Learners write the new vocabulary word in the oval. Then, they complete the information in each of the rectangle sections. By having to go beyond the dictionary definition, learners develop a deeper connection to the vocabulary and a deeper understanding of the concepts.
When to use it	Frayer models can be used after new vocabulary has been introduced or as a pre-teaching tool. They are most effective once learners have had some experience with the vocabulary in context, as those experiences help learners to determine characteristics, examples, and non-examples. Frayer models can be done in class and/or as homework.

Term	Gallery Walk
What it is	A Gallery Walk is a technique for sharing ideas that have been organized in a poster, 3D model, or other highly-visual format. A variation of the Gallery Walk is the Graffiti Wall.
How it works	Learners, often in groups, create posters or 3D models in response to a question. These models are displayed, either on walls or on tables. All learners then proceed in an orderly fashion around the room, paying close attention to the work of other groups. The debrief phase of the lesson usually asks learners to summarize what they noticed during the Gallery Walk. Sometimes, learners use sticky notes to ask questions or leave feedback for the model's creator(s). In the Graffiti Wall variation, learners leave feedback or add new information directly on the original model.
When to use it	Use a Gallery Walk when learners are ready to share ideas about concepts and need a movement break.

Term	Jigsaw (sometimes called "Numbered Heads Together")
What it is	A Gallery Walk is a technique for sharing ideas and work among groups
How it works	Each group is assigned a particular component of a larger task. Group members complete the assigned component and are then reshuffled into new groups. These groups are comprised of members from each of the original groups. Together, they put the "jigsaw" together to complete the entire task.
When to use it	Use a Jigsaw to support learners as they work through large, complex tasks.

Term	"KWL"
What it is	"KWL" stands for "What I Know, What I Want to Know, and What I've Learned." KWLs are charts that learners use to track their understanding across a lesson. Variations include "TWD" (Think, Wonder, Learned) and "KWHL" (Know, Want to Know, How I Plan to Find Out, and Learned)
How it works	At the beginning of a lesson, learners complete the "K" section by listing out everything they believe to be true about a topic. Next, learners add their questions about the topic in the "W" section. The "W" section can also be used to list conflicting ideas that the class might have about a topic. Once the "W" section is complete, learners begin researching the topic through the lesson activities. Students add their new or clarified knowledge to the "L" section at the end of the lesson. During the lesson debrief at the end, students compare their ideas in "K" to their understandings in "L." They also check to see which questions were answered and which remain as topics for future investigation.
When to use it	KWL charts can be used to see what students know (or think they know) when beginning a new topic of study. They are also useful when students are going to use print or digital materials to research a topic.

Term	Stay-Stray
What it is	Stay-Stray is a technique used for sharing ideas among groups. It is similar to a Gallery Walk, but the focus is on conversation.
How it works	When it is time for groups to share their work, one group member is designated as the "speaker." The speaker stays at the group's assigned work space and is responsible for explaining the group's ideas. The other group members are designated as "strayers." Strayers are responsible for visiting the other groups and learning how they approached their work. After a few minutes, all group members return to their assigned work areas. Strayers share with their group members what they learned by visiting the other stations. The speaker shares any questions or new ideas that came up in conversation with strayers from other groups.
When to use it	Use Stay-Stray as a tool for promoting conversation among learners while sharing group work results. It is especially useful when learners have been sitting and need a movement break.

Term	Think Time
What it is	Think Time is just what it says: time given to learners to think. "Think-Pair-Share," "Think-Write-Pair-Share," "Turn and Talk," and "Wait Time" are all common variations of Think Time.
How it works	After asking a question, wait at least 5 seconds before collecting student responses. This is "Wait Time." Think Time can be extended by having learners jot ideas down and/or share their responses with a partner ("Think-Pair-Share" or "Turn and Talk") prior to sharing with the whole group. Providing Think Time is important because it allows learners time to gather their thoughts and mentally practice their responses. It also allows learners to think more deeply about questions, opening up opportunities for richer discussion. Finally, it holds more learners accountable. All learners are more likely to think about the question when they know that "the answer" won't be shared immediately. Using partner talk increases the accountability even more, as all learners then need to participate in the lesson.
When to use it	Every time a question is posed.

Term	Ticket Out the Door (also known as "Exit Ticket")
What it is	Ticket Out the Door is a quick (2-3 minute) task that learners complete at the very end of the lesson.
How it works	At the very end of a lesson, learners complete a quick task, no more than 2-3 minutes in length. Common tasks include application questions that learners answer based on the lesson, sketching activities, or reflective activities. Sometimes, Tickets Out the Door ask learners to write down areas of confusion. Tickets Out the Door provide immediate, formative feedback for teachers to help plan for the next lesson. They allow teachers to see what learners have taken away from the lesson as well as what items are still confusing and need further clarification.
When to use it	Tickets Out the Door can be used when new material is introduced or learners encounter particularly difficult material in a unit. Use them any time more information is needed about learner progress.

Appendix D

GLOSSARY OF SCIENTIFIC TERMS

Appendix D: Glossary of Scientific Terms

In this section, you will find definitions for the scientific vocabulary terms that appear throughout the unit. If you have questions about terms that refer to teaching activities, please see the section entitled, "Teaching Tips and Techniques."

Adaptive immunity: distinguishes between substances made by the organism (self) and substances that are not part of the organism (non-self); includes antibody proteins and other proteins that recognize, bind to, and aid in the destruction of a bacteria or virus

Aerobes: organisms that cannot live without oxygen

Agar: nutritious media for microbes to grow on

Anaerobes: organisms that can live without oxygen

Antibodies: proteins that recognize and bind to antigens and destroy pathogens

Antigens: protein molecules on the surfaces of viruses, bacteria, and other pathogens

Bacteria: a type of single-celled organism. Bacteria are classified as prokaryotic organisms, as they do not have membrane-enclosed nuclei or other similar structures (organelles). Bacteria are the most abundant life forms on our planet. Singular: bacterium

Binary fission: the process of a bacterium splitting and becoming two

Biodiversity: the variety of species that inhabit an ecosystem

Cell Viability: What percentage of a sample of organisms are alive

Chill: keep food at or below 40 degrees Fahrenheit or frozen until ready to cook

Colony: collection of the same type of microbes that grow from a single cell

Concentration: measurement for the strength of a solution

Contamination: the act of contaminating, or of making something impure or unsuitable by contact with something unclean, bad, etc.

Cross contamination: the passing of bacteria, microorganisms, or other harmful substances indirectly from one patient to another through improper or unsterile equipment, procedures, or products

Defrost: to thaw or partially thaw

Dilute: the act of making a solution less concentrated

DNA pairing: bases on one strand are paired specifically with a base on the opposite strand – A always pairs with T and C with G

DNA sequences: each DNA sequence is unique to an individual organism and can be used for identification

DNA structure: is a double strand with a sequence of four bases along the molecule. Each base on one strand is paired specifically with a base on the opposite strand – A always pairs with T and C with G

- Information is contained in the order of bases along the DNA molecule
- Molecular tools can be used to identify the sequence of DNA molecules
- DNA sequences are unique to individual organisms and can be used for identification

Epidemiology: the branch of medicine dealing with the incidence and prevalence of disease in large populations and with detection of the source and cause of epidemics of infectious disease

Facultative anaerobe: organisms that can live with or without oxygen

First line of defense: part of the innate immune system that includes barriers such as the skin, tears, mucus, and low pH

Foodborne illness: food poisoning

Gel electrophoresis: a technique for separating protein molecules of varying sizes in a mixture by moving them through a block of gel, as of agarose or polyacrylamide, by means of an electric field,with smaller molecules moving faster and therefore farther than larger ones.

Growth curve: a graph of population size over time

Immune system: a network of cells, tissues, and organs that protects the human body

Innate immunity: nonspecific immunity that works against wide variety of invasive organisms

Microbiome: the term given to the microbes that live within/on a particular organism

Molarity: the SI unit for concentration

Morphology: study of the form and structure of things

Outbreak: a sudden, violent, or spontaneous occurrence, esp of disease or strife.

Pathogen: a disease-causing agent, such as a microbe, that harms the host organism

Petri Dish: common container to hold agar

Preventive measures: a variety of interventions that can be undertaken to prevent or delay the occurrence of disease or reduce further transmission or exposure to disease

Prokaryote: Any cellular organism that has no nuclear membrane, no organelles in the cytoplasm except ribosomes, and has its genetic material in the form of single continuous strands forming coils or loops, characteristic of all organisms in the kingdom Monera, as the bacteria and blue-green algae.

Second line of defense: part of the innate immune system that includes phagocytic cells which ingest foreign cells and other particles

Serial Dilution: A process by which laboratories can make very accurate small-scale solutions

Solute: the substance dissolved in a solution

Solution: a homogeneous, molecular mixture of two or more substances

Solvent: the substance into which a solute is dissolved in a solution

Thaw: bring to a more normal temperature, especially to room temperature

Third line of defense: part of the adaptive immune system that includes antibody proteins and other proteins that recognize, bind to, and aid in destruction of bacteria or viruses

Vaccinations: Inoculation with a vaccine in order to protect against a particular disease.

REFERENCES

Lesson 1

Archer, J. 16880, *Streptococcus pneumoniae.* CDC/James Archer. [Online Image]. Retrieved January 19, 2017, from https://phil.cdc.gov/phil/details.asp Public Domain.

"Body Bacteria" article: MacLachlan, A. (n.d.). Retrieved from website: http://publications.nigms.nih.gov/findings/jan12/body-bacteria.asp

Body Bacteria graphic in Powerpoint: http://publications.nigms.nih.gov/findings/jan12/body-bacteria.asp

Carr, J. 5715, *Helicobacter* sp. *flexispira.* CDC/ Dr. Patricia Fields, Dr. Collette Fitzgerald. [Online Image]. Retrieved January 19, 2017, from https://phil.cdc.gov/phil/details.asp Public Domain.

Crowhurst, L. blue orange [Online image]. Retrieved March 13, 2017 from https://flic.kr/p/9S9M8 licensed under CC BY-ND 2.0

Eckert, A. ,Oosthuizen J. 21918, Computer-generated image of a number of *Salmonella* serotype Typhi bacteria.CDC/James Archer. [Online Image]. Retrieved January 19, 2017, from https://phil.cdc.gov/phil/details.asp Public Domain.

Literacy Skills: (n.d.). Retrieved from http://www.literacyta.com/literacy-skills

March, D. bacteria [Online image]. Retrieved March 13, 2017 from https://flic.kr/p/9vELe9 licensed under CC BY 2.0

microbiologybytes. World [Online image]. Retrieved March 13, 2017 from https://flic.kr/p/wiiEsr licensed under CC BY-SA 2.0

Osman, I. Anaerobes [Online image]. Retrieved March 13, 2017 from https://flic.kr/p/gUWVWU licensed under CC BY 2.0

Osman, I. bacillus [Online image]. Retrieved March 13, 2017 from https://flic.kr/p/fHW4qF licensed under CC BY 2.0

Waters, S. prokaryote [Online image]. Retrieved March 13, 2017 from https://flic.kr/p/5R8ogz licensed under CC BY-SA 2.0

Lesson 1 Print Materials

Archer, J. 16786, *Clostridium difficile.* CDC/James Archer. [Online Image]. Retrieved January 19, 2017, from https://phil.cdc.gov/phil/details.asp Public Domain.

Archer, J. 16880, *Streptococcus pneumoniae.* CDC/James Archer. [Online Image]. Retrieved January 19, 2017, from https://phil.cdc.gov/phil/details.asp Public Domain.

Carr, J.H. 11846, *Sebaldella termitidis* bacteria. CDC/ Brian J. Beck, PhD, American Type Culture Collection (ATCC) [Online Image]. Retrieved January 19, 2017, from https://phil.cdc.gov/phil/quicksearch.asp. Public Domain.

Carr, J. 5715, *Helicobacter* sp. *flexispira.* CDC/ Dr. Patricia Fields, Dr. Collette Fitzgerald. [Online Image]. Retrieved January 19, 2017, from https://phil.cdc.gov/phil/details.asp Public Domain.

CDC. 15233, *Peptostreptococcus anaerobius* and *P. asaccharolyticus (*formerly *Peptococcus asaccharolyticus).* CDC/ Jim Pledger, Dr. V.R. Dowell, Jr. [Online Image].
Retrieved January 19, 2017, from https://phil.cdc.gov/phil/details.asp Public Domain.

Eckert, A. ,Oosthuizen J. 21918, Computer-generated image of a number of *Salmonella* serotype Typhi bacteria.CDC/James Archer. [Online Image].
Retrieved January 19, 2017, from https://phil.cdc.gov/phil/details.asp Public Domain.

Jones, G.L.. 3073, *Bifidobacterium eriksonii.* CDC/Gilda L. Jones [Online Image].
Retrieved January 19, 2017, from https://phil.cdc.gov/phil/details.asp Public Domain.

NIAID. E. coli Bacteria [Online Image].
Retrieved December 14, 2016 from www.flickr.com licensed under CC BY 2.0

NIAID. Methicillin-Resistant Staphylococcus aureus (MRSA) [Online Image].
Retrieved January 19, 2017, from www.flickr.com licensed under CC BY 2.0

NIAID. 18132, *Borrelia burgdorferi* bacteria. National Institute of Allergy and Infectious Diseases (NIAID). [Online Image].
Retrieved January 19, 2017, from https://phil.cdc.gov/phil/details.asp Public Domain.

Strong, B. 3079, *Bifidobacterium eriksonii* bacteria. CDC/ Bobby Strong. [Online Image].
Retrieved January 19, 2017, from https://phil.cdc.gov/phil/details.asp Public Domain.

Wardell, P. 12803, *Enterococcus faecalis* sp. Centers for Disease Control and Prevention. [Online Image].
Retrieved January 18, 2017, from https://phil.cdc.gov/phil/details.asp Public Domain.

Lesson 2A

Gathany, J. Sneeze in progress. CDC/ Brian Judd [Online image].
Retrieved March 13, 2017 from https://phil.cdc.gov/phil/details.asp?pid=11162 Public Domain

Howard, L. Scanning electron microscope image of lung trachea epithelium [Online image].
Retrieved March 13, 2017 from https://commons.wikimedia.org/wiki/File%3ABronchiolar_epithelium_4_-_SEM.jpg
Public Domain

Scottow, A. Skin [Online image].
Retrieved March 13, 2017 from https://flic.kr/p/dwkMY8 licensed under CC BY-SA 2.0

zenjazzygeek. Donkey [Online image].
Retrieved March 13, 2017 from https://flic.kr/p/8AM8Jp licensed under CC BY 2.0

Lesson 2B

Mango Slices. Phagocytosis [Online image].
Retrieved March 13, 2017 from https://commons.wikimedia.org/wiki/File%3APhagocytosis.svg
licensed under CC BY-SA 4.0

NIAID. Healthy Human T Cell [Online image].
Retrieved March 14, 2017 from https://flic.kr/p/a4RLoY licensed under CC BY 2.0

NIAID. Immune Cells Surrounding Hair Follicles in Mouse Skin [Online image].
Retrieved March 14, 2017 from https://flic.kr/p/cNzEzb licensed under CC BY 2.0

Rosen, Y. Actinomycosis [Online image].
Retrieved March 14, 2017 from https://flic.kr/p/94dPwV licensed under CC BY-SA 2.0

Lesson 2 Print Materials

NIAID. Human B Lymphocyte [Online Image].
Retrieved January 26, 2017, from www.flickr.com licensed under CC BY 2.0

NIAID. Colorized scanning electron micrograph of a macrophage. [Online Image].
Retrieved January 26, 2017, from www.flickr.com licensed under CC BY 2.0

Lesson 3

CDC 20581, Chinese epidemiologists. CDC/ Jessie Blount. [Online Image].
Retrieved January 24, 2017, from https://phil.cdc.gov/phil/details.asp Public Domain.

CDC 19903, Field Epidemiology Training Program (FETP) resident. CDC. [Online Image].
Retrieved January 24, 2017, from https://phil.cdc.gov/phil/details.asp Public Domain.

"epidemiology". Dictionary.com Unabridged. Random House, Inc. 23 Jan. 2017. <Dictionary.com http://www.dictionary.com/browse/epidemiology>.

Laserson, K. 19838, Epidemic Intelligence Service (EIS) officer. CDC/ Dr. Kayla Laserson, India. [Online Image].
Retrieved January 24, 2017, from https://phil.cdc.gov/phil/details.asp Public Domain.

"outbreak". Collins English Dictionary - Complete & Unabridged 10th Edition. HarperCollins Publishers. 23 Jan. 2017. <Dictionary.com http://www.dictionary.com/browse/outbreak>.

"vaccination". The American Heritage® Science Dictionary. Houghton Mifflin Company. 23 Jan. 2017. <Dictionary.com http://www.dictionary.com/browse/vaccination>.

Lesson 3 Print Materials

Goetz, G. "Salmonella Outbreak Linked to N.C. Holiday Inn Sickened 100, Final Report Says." Food Safety News.
Retrieved January 24, 2017. http://www.foodsafetynews.com/2013/07/salmonella-outbreak-linked-to-north-carolina-holiday-sickened-100-final-report-says/#.WIfDprYrlhs. Used with permission.

Lesson 4

Genome.gov. DNA structure and bases [Online image].
Retrieved March 14, 2017 from https://commons.wikimedia.org/w/index.php?curid=510095 Public Domain

Hull, D. DNA / Protein function finder [Online image].
Retrieved March 14, 2017 from https://flic.kr/p/7JQMHY licensed under CC BY 2.0

Hull, D. Genome Trumps (Cat) [Online image].
Retrieved March 14, 2017 from https://flic.kr/p/7JLS8H licensed under CC BY 2.0

Lakdawalla, A. Radioactive Flourescent Seq [Online image].
Retrieved March 14, 2017 from https://commons.wikimedia.org/wiki/File:Radioactive_Fluorescent_Seq.jpg#/media/File:Radioactive_Fluorescent_Seq.jpg licensed under CC BY-SA 3.0

National Human Genome Research Institute. Next Generation DNA Sequencing [Online image].
Retrieved March 14, 2017 from https://flic.kr/p/Js2mU6 licensed under CC BY 2.0

US Department of Agriculture. k5965-1 [Online image].
Retrieved March 14, 2017 from https://flic.kr/p/dTdqaR licensed under CC BY 2.0

Lesson 4 Print Materials

Loris. Sanger sequencing read display. Commons.wikimedia.org. [Online Image].
Retrieved January 25, 2017, from https://commons.wikimedia.org/wiki/File:Sanger_sequencing_read_display.png
Public Domain.

Lesson 5

Cann, A. J. Lactobacillus casei [Online image].
Retrieved March 14, 2017 from https://flic.kr/p/dHofYV licenses under CC BY-SA 2.0

Carr, J.H. 10047, Scanning electron microscopic view of MRSA. CDC/Jeff Hageman, M.H.S. [Online Image].
Retrieved January 31, 2017, from https://phil.cdc.gov/phil/details.asp Public Domain.

cnicholsonpath. Helicobacter heilmanii [Online image].
Retrieved March 14, 2017 from https://flic.kr/p/bxpcjY licensed under CC BY 2.0

Dumb Scientist. Binary_Fission [Online image].
Retrieved March 14, 2017 from https://flic.kr/p/8KE5QH licensed under CC BY-SA 2.0

Eppink, J. Danger: Oxygen [Online image].
Retrieved March 14, 2017 from https://flic.kr/p/d5xmAL licensed under CC BY 2.0

Francisco, M. Escherichia coli and Staph. aureus [Online image].
Retrieved March 14, 2017 from https://flic.kr/p/aUqzZR licensed under CC BY 2.0

Joel, J. 141087-050-24850517 [Online image].
Retrieved March 14, 2017 from https://flic.kr/p/fbxaHJ licensed under CC BY-ND 2.0

NIAID. Salmonella Bacteria [Online image].
Retrieved March 14, 2017 from https://flic.kr/p/9y4sD6 licensed under CC BY 2.0

NIAID. Staphylococcus epidermidis Bacteria [Online image].
Retrieved March 14, 2017 from https://flic.kr/p/9y68Ts licensed under CC BY 2.0

Lesson 5 Print Materials

Carr, J. 5715, *Helicobacter* sp. *flexispira.* CDC/ Dr. Patricia Fields, Dr. Collette Fitzgerald. [Online Image].
Retrieved January 19, 2017, from https://phil.cdc.gov/phil/details.asp Public Domain.

Eckert, A. ,Oosthuizen J. 21918, Computer-generated image of a number of *Salmonella* serotype
Typhi bacteria.CDC/James Archer. [Online Image].
Retrieved January 19, 2017, from https://phil.cdc.gov/phil/details.asp Public Domain.

Health Gauge. Sugar Spoon [Online Image].
Retrieved February 3, 2017 from www.flickr.com licensed under CC BY 2.0

NIAID. E. coli Bacteria [Online Image].
Retrieved December 14, 2016 from www.flickr.com licensed under CC BY 2.0

NIAID. Methicillin-Resistant Staphylococcus aureus (MRSA) [Online Image].
Retrieved January 19, 2017, from www.flickr.com licensed under CC BY 2.0

Polgar, E. Bodie, United States (desert, wagon, boarded-up house). [Online Image]. Retrieved February 3, 2017, from https://unsplash.com/licensed under CC 0 1.0.

Strong, B. 3079, *Bifidobacterium eriksonii* bacteria. CDC/ Bobby Strong. [Online Image]. Retrieved January 19, 2017, from https://phil.cdc.gov/phil/details.asp Public Domain.

Vladivostok, M. Cefazolin ball-and-stick.png. [Online Image]. Retrieved February 3, 2017, from https://commons.wikimedia.org/licensed under CC 0 1.0.

Vladivostok, M. Ball-and-stick model of Amikacin molecule. [Online Image]. Retrieved February 3, 2017, from https://commons.wikimedia.org/licensed under CC 0 1.0.

Windus, C. Rusty Car. [Online Image]. Retrieved February 3, 2017, from https://unsplash.com/ licensed under CC 0 1.0.

Lesson 6

CDC. 19760, Haiti's National Public Health Lab, training laboratory students. [Online Image]. Retrieved March 13, 2017, from https://phil.cdc.gov/phil/details.asp Public Domain.

Ismael, W. Sea Salt Water for real Hydration [Online image]. Retrieved March 15, 2017 from https://flic.kr/p/nLSegU licensed under CC BY-ND 2.0

Dollete, R. Sweet Tea [Online image]. Retrieved March 15, 2017 from https://flic.kr/p/9kcBh1 licensed under CC BY-ND 2.0

Andi. Mole [Online image]. Retrieved March 15, 2017 from https://flic.kr/p/5NynH licensed under CC BY-SA 2.0

News 21 – National. Working with Salmonella [Online image]. Retrieved March 15, 2017 from https://flic.kr/p/9ZYfDw licensed under CC BY 2.0

Grab and Go 2

CDC. 11765, Petri dish innoculated with *Francisella tularensis.* CDC/ Megan Mathias and J. Todd Parker [Online Image]. Retrieved March 13, 2017, from https://phil.cdc.gov/phil/details.asp Public Domain.

Grab and Go 3

Andrade, E. George's Hamburgers [Online image]. Retrieved March 14, 2017 from https://flic.kr/p/r1PeG licensed under CC BY-ND 2.0

Bittinger, N. I am NOT eating dough! [Online image]. Retrieved March 14, 2017 from https://flic.kr/p/5JD8gj licensed under CC BY 2.0

Fight Bac! 4Quad_LOGO [Online image]. Retrieved March 14, 2017, from http://www.fightbac.org/wp-content/uploads/2015/08/4Quad_LOGO.jpg. Used by permission.

Fight Bac! Be_Food_Safe_logo gif [Online image]. Retrieved March 14, 2017, from http://www.fightbac.org/wp-content/uploads/2015/08/Be_Food_Safe_logo.gif. Used by permission.

Fight Bac! Be_Food_Safe_logo jpg [Online image].
Retrieved March 14, 2017, from http://www.fightbac.org/wp-content/uploads/2015/08/Be_Food_Safe_logo.jpg.
Used by permission.

Fight Bac! logo_chill [Online image].
Retrieved March 14, 2017, from http://www.fightbac.org/wp-content/uploads/2015/08/logo_chill.gif.
Used by permission.

Fight Bac! logo_clean_produce [Online image].
Retrieved March 14, 2017, from http://www.fightbac.org/wp-content/uploads/2015/08/logo_clean_produce.gif. Used by permission.

Fight Bac! logo_separate [Online image].
Retrieved March 14, 2017, from http://www.fightbac.org/wp-content/uploads/2015/08/logo_separate.gif.
Used by permission.

Mills, A. 13578, Washing a red pepper. CDC/ Amanda Mills [Online Image].
Retrieved March 13, 2017, from https://phil.cdc.gov/phil/details.asp Public Domain.

Grab and Go 3 Print Materials

Fight BAC! Brochure from The Partnership for Food Safety Education-used with permission. [Online PDF].
Retrieved March 7, 2017, from http://www.fightbac.org/free-resources/brochures-flyers/

www.ingramcontent.com/pod-product-compliance
Lightning Source LLC
LaVergne TN
LVHW070910210926
848482LV00017B/1041